PLEASE, MR. EINSTEIN

Jean-Claude Carrière

PLEASE, MR. EINSTEIN

Translated from the French by
John Brownjohn

HARCOURT, INC.

Orlando   Austin   New York   San Diego   Toronto   London

Requests for permission to make copies of any part of the work
should be submitted online at www.harcourt.com/contact or mailed to
the following address: Permissions Department, Harcourt, Inc.,
6277 Sea Harbor Drive, Orlando, Florida 32887-6777.

www.HarcourtBooks.com

This is a translation of *Einstein S'il Vous Plaît.*
First published in English in Great Britain by Harvill Secker.

This book is supported by the French Ministry of Foreign Affairs, as part of
the Burgess Programme headed for the French Embassy in London by the
Institut Français du Royaume-Uni.

Library of Congress Cataloging-in-Publication Data
Carrière, Jean-Claude, 1931–
[Einstein, s'il vous plaît. English]
Please, Mr. Einstein/Jean-Claude Carrière;
translated from the French by John Brownjohn.
p.   cm.
I. Brownjohn, John.   II. Title.
PQ2663.A78E3613   2006
843'.914—dc22   2006006361
ISBN-13: 978-0-15-101422-4   ISBN-10: 0-15-101422-1

Text set in Adobe Garamond
Designed by Linda Lockowitz

Printed in the United States of America
First U.S. edition
A C E G I K J H F D B

PLEASE, MR. EINSTEIN

- ONE -

LET'S FOLLOW THAT GIRL who's walking down the street. She waits for some cars to go by, then crosses without bothering about the lights.

We're in a Central European city—Prague or Vienna, perhaps, or Munich or Zurich. There's no well-known monument in sight that would enable us to identify the city in question. The weather is quite fine, the time of year indeterminate. It's late afternoon and the shadows are beginning to lengthen. The girl is wearing jeans, flat-heeled black shoes and a blouse. Aged between twenty and twenty-five, she's on the slim side, with an animated face and brisk movements. She's carrying a shoulder bag. We might take her for a student, but a student in her final year.

This, then, is the moment at which we first catch sight of her. We will never know where she comes from, or what her name is, or what her parents do, or how her life will turn out. We're following her simply because our gaze has lighted on her in the street.

She hears the bell of an approaching tram and skips onto the opposite pavement. The yellow and black tram, which she has failed to see or hear until now, narrowly misses her. It bears the number 17.

She watches it recede, then looks up. Above her she sees a building dating from 1910 or 1920, with a ponderous, rather dreary façade. She takes a crumpled piece of paper from one of the pockets in her jeans and checks the address on it.

Yes, this is the place all right. She goes in.

There's nothing remarkable about the entrance hall. She makes her way along it and up a shadowy flight of stairs whose varnished treads, partly covered by a narrow strip of beige carpet, creak in places. Running her hand up the rather chunky wooden banister rail, she quickly climbs the stairs to the first floor. There, after peering into the gloom for a moment or two, she rings a doorbell.

We don't know exactly what time it is, but in any case, the girl seems quite unconcerned whether it's morning or evening, Monday or Wednesday.

She settles down to wait on the landing, but the door opens almost at once. It's held ajar by a dark, elderly woman in a longish skirt and an old-fashioned blouse trimmed with lace. We can't tell if the lace is handmade (though it's possible).

The woman has a strong, calm face with a prominent nose and pale skin displaying only faint traces of makeup— a kindly enough face for someone who's answering a door.

She asks the visitor if she's expected. Does she have an appointment?

"Not really," the girl replies. "I had some time to spare, so I came just like that—on the off chance, so to speak. I can come back if necessary. Or wait for as long as I have to."

"You're sure you've come to the right address?"

"I think so."

She holds out the piece of paper. The pale-faced woman glances at it. She hesitates briefly, very briefly, then opens the door a little wider and steps aside.

"Come in," she says grudgingly.

"Thanks."

The girl squeezes into the apartment. We follow her.

After making her way through a kind of recess she finds herself in a windowless waiting room where a dozen people, men for the most part, are patiently seated on some nondescript chairs, not all of the same design.

They glance up at the new arrival. As for her, she surveys the room with interest but little surprise before sitting down on the only unoccupied chair. Some of the men are wearing clothes and shoes that look as if they date from the first half of the twentieth century, or the 1950s at the latest. They've all taken the trouble to put on ties, though one or two of their rather grubby shirt collars are curling up rebelliously. Their jackets are buttoned. Nearly all of them are holding leather briefcases or bulky folders under their arms or on their laps. Reposing on some of these folders, most of them firmly secured with straps, are felt hats.

Out of the corner of her eye, the girl notes that the majority of the people waiting here are clutching their briefcases and folders tightly, even to the extent of digging their nails into them as if they contain fortunes in paper.

One of them gives a start whenever he hears the muffled hiccups of the central heating system buried somewhere in the old building's pipework.

Also audible, and emanating from the street, is the sound of trams passing in both directions with bells clanging. But that startles no one. It's like a punctuation mark, an urban cadenza.

One of the waiting men, who had deposited his black briefcase on the floor, propped against the legs of his chair, bent down and retrieved it when the girl came in. As if suddenly afraid of something, some indiscretion or attempt to steal it, he's now clasping it to his chest with both hands.

Another man, seated in the waiting room's only armchair, is a stern-looking individual in a gray wig, quite a long, curly one, which he doesn't attempt to disguise. He's wearing a kind of voluminous, old-fashioned cloak over his street clothes, which appear to be dark, and shoes with silver buckles.

The girl notices these unwonted details without seeming too surprised to find herself with such people in such surroundings. Was she expecting it? We can't tell, not being privy to her thoughts. In any case, she isn't intimidated. She glances at her watch, glances at it again, then peers more closely and shakes her wrist as if it has stopped. She looks round inquiringly, but there's no clock on the wall or the mantel.

She turns to the man sitting beside her and asks him the time in a low voice.

"I don't know what time it is," he replies in a Central European accent.

"But roughly?"

"No, I'm sorry, I don't know."

She catches the eye of a square-faced, white-haired woman who doesn't wait to be asked. "Nor do I," says the woman.

"It's late, anyway," the Central European sees fit to add.

The square-faced woman nods. She also thinks it's late.

We sense that the girl is gradually becoming disconcerted, almost perturbed, by this whole setup. She seemed on arrival to be quite relaxed—proof against surprise and emotion of any kind. She doesn't look as if she's come here for medical treatment or legal advice. She might be in the outer office of a theater manager or casting director, waiting with other hopefuls to be selected for a part in a film or a play. But in that case, why are all these people around her clutching briefcases?

In any event, if she's after a part she doesn't have competition on this particular day. She's the only girl sitting there.

She seems doubly surprised when the woman in the long skirt—let's call her Helen, which we later discover is her name—reappears, points to her and beckons her into another room.

"You. Come with me, please."

"Me?"

"Yes, you. Come along."

It's possible that the girl feels tempted to say that there's been some mistake, that she was the last to arrive and has plenty of time to spare. We often hesitate at the last moment when our turn comes and the time for a decision approaches. We would prefer to prolong the tedium of waiting—to remain in ignorance of a diagnosis, for example, when we go to the doctor. In the end, however, the girl says nothing and complies. For one thing, she has no idea of the rules that must prevail in this waiting room, this building. For another, she finds it very hard to tell how long she's been sitting here. Like her, we're under the impression that she's only just come in, but that impression lacks certainty and precision, and there's no clock or watch to corroborate it. Perhaps she has been waiting on this chair for longer than she thinks, longer than she seems to have been waiting. It doesn't matter.

Under the resentful gaze of the others, who have doubtless been waiting for ages (we will never know how long) and who suddenly see this latecomer cut the line without explanation, the girl gets up, crosses the room with her bag on her shoulder and follows Helen into an adjacent room.

The door closes behind the two women. Some of the waiting men sigh, others mutter irritably to themselves. The one who snatched up his briefcase replaces it gently on the floor. Another clears his throat and coughs a couple of times.

A tram goes by in the street. The bell clangs twice.

THE GIRL MAKES HER WAY through some double doors. She now finds herself in a sizable study whose contents include books, periodicals and pamphlets, various documents and instruments, a blackboard (complete with rag and chalk), a violin in its case, some scattered sheets of music and a music stand. Here too, the furniture looks as if it dates from the beginning or middle of the twentieth century.

She quickly surveys her new surroundings without being able to take note of every last detail in the few seconds available.

She does, however, see that three other doors, at present shut, give access to the room. Three doors in addition to the one by which she has entered it.

Behind her she hears a man's rather high-pitched, almost quavering voice.

"That'll be all, Helen."

She turns to find herself face to face with Albert Einstein, whose appearance has scarcely been touched by eternity.

Yes, it's Albert Einstein in person. There's no mistaking him. He looks about fifty-five or sixty, with the thick mustache and long, almost white hair we know from photographs, the dark eyes and heavy lids drooping at the corners, the high, furrowed brow. The face of a man known the world over.

Very simply dressed in a rather crumpled pair of slacks, a shabby, pale beige sweater and leather sandals worn without socks, he's smiling and rather awkward in his movements.

Helen withdraws by way of a small door hidden behind a bookcase—a door the girl failed to notice when she came in. This brings the number of doors to five.

"See you later, sir," Helen says to Einstein before closing the hidden door behind her. "I'll be here if you need me."

"Very good," Einstein replies. "Thank you, Helen."

"What should I do about the others?" she adds, holding the fifth door ajar.

"Oh, we'll see."

"Are they to wait?"

"Yes, have them wait."

Now alone with Albert Einstein in his study, the girl gives him a little nod. She's looking a trifle intimidated, and we can well understand why. However, since she appears to be pretty self-possessed or even bold by nature, she quickly recovers her composure under the celebrated scientist's gaze.

"You said that time doesn't exist," she tells him, "so I took the liberty of coming to see you."

"You did the right thing," Einstein replies without taking his eyes off her.

"I took a chance."

"One can be lucky sometimes."

"I didn't think I would be, but here I am."

Possibly for want of something else to say, she adds, "My watch has stopped."

"It happens, especially in this building. Everyone complains about it, I gather. Personally, I've given up wearing one."

He suddenly bursts out laughing. It's a resounding, booming peal of laughter, as if he'd just uttered some hilarious remark. His reaction rather disconcerts the girl.

She waits for him to stop laughing, then: "And you don't miss it?"

"What?"

"Your watch."

This time he smiles.

"Oh, not at all," he replies with a little shrug. "Least of all now. I mean, what would I do with a watch?"

They look at each other in silence. For how long? How can we judge? They don't know what to say or do. Each seems to be waiting for the other to make some signal. The girl notices that Einstein is examining her closely, observing her face, her hands, her figure, her clothes, her shoulder bag. If she has read any books about him, which is probable (since it's he whom she came to see in this building and his presence there didn't surprise her, even though we don't know where she obtained her information), she knows that he has never been indifferent to the opposite sex—far from it—and that there was a time when the women in his life succeeded one another in pretty short order. She also knows that most men find her extremely attractive. They even contrive to tell her so, or to intimate as much.

Has she decided to gamble on this? We can't tell, not yet. She probably doesn't even know herself.

At this stage it's futile for us to speculate about her true intentions, about what she expects and is thinking or hoping. Futile and factitious. We don't know her well enough for that. We don't even know what preparations she made for this visit or what she really hopes to gain from it. We followed her along the street and up the stairs; that's pretty much all we know about her.

If instead of reading a book—or writing it—we were getting acquainted with these characters and this situation in a cinema or in front of a television screen, or even at the theater, we wouldn't ask ourselves any subsidiary questions, and for a very good reason: we wouldn't have time. We would be so thoroughly carried along by the action of the film or play—provided the action was sufficiently appealing and engrossing, or at best irresistible—that this would stifle any kind of inquiry into the attendant circumstances.

It would be impossible for us to see a particular episode again, to replay it and check on what has just been written or filmed; we could only watch and listen. As in the present instance.

Doubtless casting around for something to say, the girl gestures toward the door.

"There are lots of people waiting outside."

"I can't help that."

"Is that your waiting room?"

"If you like, but I'm not a lawyer or a dentist, you know."

"They come to consult you, though?"

"Yes, in a manner of speaking. Consult me? Yes, sometimes."

"What about?"

"Oh, all sorts of things. The design of a new electric iron, a printing press, a motor, an optic fiber. About what goes on in the heart of a star or a comet's tail. But most of the time they come to demonstrate, armed with supporting evidence, that I was utterly and completely wrong."

"And it's like that every day?"

"Yes, it's always like that. Or so it seems to me, anyway. But I must confess I've more or less forgotten what 'every day' means. I've given up using a calendar. It's the same with a watch. They're ideas that have eluded me."

"Where do they come from, those people out there?"

"All over the place. I don't ask them where or when they hail from."

"They look as if they're from another age."

"That's because they've been waiting a long time, wouldn't you think?"

The girl doesn't speak for a while—exactly how long, it's as hard to determine as before. She looks around Einstein's study, then back at Einstein himself—Einstein, who has been dead for fifty years (all the reference books say so), but who is there in front of her, apparently in the flesh, in a study that seems to be his and is visited by all those oddly attired people.

"But you saw me first," she says.

"Really?"

"Does that surprise you? Wasn't the decision yours?"

"No. I never decide such things."

"Who does? The woman I saw just now?"

"Helen? Yes, of course. It was she who told me about you. A girl has just turned up, she said."

"Is that what she told you? So why did you let me in?"

He spreads his arms, palms upwards.

"Because I'm very glad to see you," he replies, still smiling.

"Why are you glad to see me?"

His smile quickly vanishes. He hesitates for a moment or two.

"Because," he says at length, "since you apparently come from what you still call the future, your presence here proves to me that the human race hasn't disappeared."

"You were afraid it had?"

"Very much so!"

"Because of the atomic bomb?"

He cocks his right forefinger.

"Because of nuclear weapons, to give them their correct designation."

"If the human race *had* disappeared, would you be afraid of being held responsible?"

For the first time, Einstein's expression suddenly darkens. As the facile cliché has it, a shadow flits across his face—a shadow across the face of a shade. Right at the very outset, his visitor has boldly asked the question that troubles him most.

"Yes," he replies in a subdued voice. "Yes, it's true. I might be afraid of that."

"But responsible to whom? If humanity had disappeared, who would be left to blame you for it?"

Einstein's smile returns. He beams at this notion, seemingly appreciative of his visitor's rapid train of thought.

"No one, you're right. But all the same . . ."

They preserve a thoughtful silence, possibly envisioning, each in their own way and with their own mental resources, the irrevocable disappearance that haunts the living from birth, or at least from the moment they become aware of their own existence. Einstein gently points out that at the girl's age—twenty-two, twenty-five?—it is rare, if not unlikely or impossible, for some-one to conceive of nothingness. It's an invaluable concept, he says, but one that generally comes with age alone. To the young, all is in forward motion, in being. Existence still prevails over its antithesis, which takes root in us only gradually, like a stowaway once he's certain that we're heading straight for nothingness, that we won't escape it, that our voyage to the void, to the end of things and the end of our memory of things, may terminate tomorrow morning—even, perhaps, tonight.

The girl expresses surprise that in his present state, which she gives no thought to defining (the man facing her and looking at her makes no allusion to his own, apparently real, presence in this room), Einstein is still interested in the old dialogue between what does and does not exist.

"Oh, I'm not really interested in it anymore," he says. "It's just that I haven't forgotten it."

There may be some questions on the tip of the girl's tongue—What is he made of? What sort of being is he?—but she refrains from asking them. Each of those questions is probably too simple and ill-adapted to present circumstances. She entered the building, climbed the gloomy staircase and walked in without waiting, admitted by the woman named Helen. Everything happened in a natural, effortless, uninhibited way, without the need to answer any formidable riddles posed by some sphinx. It was as if the normal constraints of time, the minor delays, the wasted journeys, the lapses of memory, the mistakes, the rites of passage (inquiries as to one's identity and intentions) had suddenly evaporated; as if every action were becoming fluid, easy and unsubjected to the recurrent checks we all experience in our various lives.

She has broken into the realm of the impossible like a discreet burglar. It's as if the very role of time had changed—as if it were for once submitting itself, of its own accord and with good grace, to the simple force of an idea.

It's the same with certain plays whose power of persuasion is such that, despite their fundamental artifice, and despite the visible presence of scenery and actors, we end by believing in the reality of what we're seeing onstage although we know it to be an invention, a fabrication. When defining those privileged moments we speak of a "suspension of disbelief," which indicates full and total absorption. We discard our disbelief, or rather, we parenthesize it and place it "in suspense," smitten by a superior reality and even by a superior truth, a dramatic truth more powerful at such moments than any other.

After which disbelief returns, of course, because it cleaves to our skeptical substance. If it does not return—if we remain convinced of having encountered the truth—we hasten to enroll ourselves as quickly as possible in the sect to which we owe that illusion. And then we're lost.

THE YOUTHFUL VISITOR continues to survey her surroundings for a while, and it seems to her, although she isn't really sure, that certain objects have been replaced by others since she came in; that there's something vague and vicissitudinous about the very configuration of the place she's in—or rather, feels she's in. But, once again, there's no question of her returning to or missing the everyday order of things. On the contrary, prompted by her own audacity to enter a dimension whose existence she never suspected, she is, we hope, enjoying the experience. Ordinary time and even space itself, perhaps, are indulging her for a while. Better not to risk breaking the spell by excessive curiosity.

"I see you're still working," she says.

"Why not? I have nothing else to do these days. I'm almost left in peace where lectures and petitions are concerned, especially now that I've ceased to be very well informed about the geopolitical situation. I also imagine I'm only here to work. Sometimes, anyway. Yes, that's what I imagine at times, because thought is like light: it's continuous and discontinuous."

"You'll have to explain what that means."

Once more he spreads his arms, then lets his hands subside onto his thighs without a sound. A familiar gesture. He sometimes resembles a character in a picture book, a benign old gentleman in a children's story—Geppetto in *Pinocchio*, for instance.

"Explain, explain . . . People are so demanding, they always want to understand everything. I'll be glad to try to explain. I often have. In any event, I've endeavored to do so all over the place. You can't assert things without attempting to give the reasons for them, but explanations aren't always enough."

"Meaning what? Please explain."

"Come, come, young lady, don't pretend not to understand, because this is the essential point. If one is to explain something to people, they must intend and want to understand. If not, one might as well be addressing a brick wall."

"I do intend and want to understand—to learn something, even, that's why I came straight to you, to your home. I'm not here to ask you to sign a petition, I'm not campaigning for anything and I have no plans to make money off of you. I just want to know a little more. However, from what people have told me and what I've read here and there, the things you say aren't simple."

"People would like things to be simple, naturally. I'd like them to be simple myself, it might make life easier for us. But in most cases simplicity would be a lie. A sham, anyway. Things—things *per se*—aren't simple. That, at least, we know for sure. We've bidden farewell to the simplicity of the universe. There's nothing more complex and disconcerting than matter."

"What are you comparing it to?"

"I'm sorry?"

"When you say there's nothing more complex than matter, what else do you have in mind? What could we compare it to?"

He smiles at her and nods as if, being a true connoisseur, he appreciates the questions she has just asked.

"Very perceptive of you. Matter and space-time are all there is. We have no basis of comparison, only our own conception of simplicity. We could compare matter only to the absence of matter, and that, strictly speaking, would be meaningless. As for space-time, what to compare it to? We ourselves are a moment in the history of matter in a particular space-time, a moment we can call consciousness. We also possess, to all appearances, what we term a 'mind.' That's what we call it. Our mind admits that it's a mind. A mind that wishes to know and understand. That's the way it's made, right?"

"Right," she says.

"But what does it mean exactly, this word 'understand'? How to explain it?"

"All the same, you're on record as stating that the most incomprehensible thing is that the universe is comprehensible."

"Did I say that?"

"Yes."

"I must have been joking. Either that or I was feeling particularly optimistic that day. Or trying to get rid of someone. We should always beware of formulas, precisely because they're based on dangerous simplicity. In any case, I should never have said 'comprehensible'—that was a minor slip—but 'accessible,' 'reducible' or 'in conformity with our equations.' Look, if you like, and since we're talking about the difficulty of explaining things, I'll give you an example. Take continuous and discontinuous, which I happened to mention just now. Do you have some idea of what those words imply?"

"Yes, I think so."

"Go on, then. I'm listening."

Readily divining that this is a kind of entrance examination, she tries to give what she hopes will be a lucid and straightforward definition of continuous and discontinuous. Needless to say, she very soon gets in a muddle. How can one imagine a force or event that is "continuous"—in other words, free from interruptions, hiatuses, deflections and changes of state—without at the same time imagining the intervals and standstills of "discontinuity," that is to say, the exact opposite of continuity but impossible to separate from it? It's as if the continuous can be thought of only with the aid of the discontinuous. And vice versa, no doubt.

Einstein seems to be taking this little test as a game—less than seriously. He cocks a finger. Possibly trying to help his visitor, he asks, "Which, in your opinion, is a circle, continuous or discontinuous?"

"It's continuous."

"And a straight line?"

"Discontinuous."

"Are you sure?"

"That's the way I see it. A straight line has two extremities, a beginning and an end."

"You're right in one sense. You share the general view, which is natural. A circle has neither a beginning nor an end. A straight line begins and ends at some point, it's bound to. So the straight line is discontinuous, whereas the circle is continuous. To our eyes, that's beyond dispute. However, we could also say the opposite."

"The opposite?"

"Or almost. We could call a straight line continuous insofar as it connects two points. A *dotted* line would be discontinuous."

"OK."

"We could also say that a straight line is limitless because we can prolong it to infinity in our imagination, whereas a circle is self-contained and limited. If the truth be told, the archetype of continuity would be a void, the immensity of space, and the archetype of discontinuity would be a speck or atom. But all right, let's stick with your own impression, it's the commonest. Besides, some philosophers and even some scientists long regarded the circle, the perfect figure, as symbolic of divine creation and the straight line as symbolic of the finitude of the species to which they assigned us."

"Our life was a straight line, you mean?"

"Yes, with a beginning and an end. We couldn't prolong it in either direction."

"So God could only operate in circles?"

"That's what men decided. God operated in circles that furnished a representation of eternity, without beginning or end. That's why the stars were seen as circumferences. Or as spheres."

"And they aren't spheres?"

"Let's not get into that, I advise against it. We'd only become involved in a very long digression. Yes, of course. Some heavenly bodies, the planets at least, seem to our eyes to be spherical. But for altogether different reasons."

"God has nothing to do with it?"

"Not that I'm aware of. As for the old conflict between the circle and the line we term straight, it lapsed into oblivion long ago—well before I arrived on the scene. The two things have now succeeded in blending and merging. Strictly speaking, we even find it quite hard to distinguish between them."

"Really?"

"Yes, because both are the fruit of our mental processes. They share the same origin. For a long time we thought in straight lines—or tried to, at least. We thought in terms of rightness and clarity, with the aid of what we call logic, reason, even geometry. When thinking we strove to proceed in a rectilinear manner, as swiftly and simply as possible, from one point to another. Our thinking developed in triangles, in squares, in rectangles, but suddenly we were invaded by the curve! By sinuosity! We were invaded and overwhelmed just when no one was expecting it! We slid into it headfirst! Even space has become curved!"

"Because of you."

"Of me?"

"Didn't you talk about curved space?"

"Oh, yes, curved space-time. But I couldn't do otherwise."

"Why not?"

"Because science isn't written about or based on sudden whims or miraculous flashes of enlightenment. It's never the work of a single individual—it presupposes an accompaniment, a support, a consensus with other scientists. It daily tackles the thousand-headed monster we call reality. And it's a tough fight, believe me. We're always in need of fresh brains. That's because we inevitably reduce everything, absolutely everything, to our own matter, to ourselves. I would go so far as to say this: we've had to learn to forget ourselves, to think 'aside,' beyond outward appearances, even beyond what is human."

"But when you refer to thought—here, at this moment—surely you mean human thought?"

"Up to now, it's the only kind we know of."

"Do you regret that?"

"Oh yes, because it would be thrilling—staggering, fantastic, extraordinary—to encounter thought processes other than our own. Far more amazing than the discovery of long lost continents, unknown plants or minerals. But alas, we can only imagine these extra-human or extraterrestrial thoughts because they all stem from our own. Even if they hypocritically claim the contrary—even if we attribute them to reddish creatures from a parallel world or to giant spiders—they're thoroughly human."

"If I understood you correctly, you said that human thought is both continuous and discontinuous."

"I've been able to ascertain that in myself. Many times."

"How?"

"It's called absent-mindedness, and all of us are prone to it. When I used to go sailing—oh, in a very modest way, on a lake in Germany—I nearly always went out on my own. I wanted to take my mind off things, as they say, so I put out from the shore thinking only of what I had to do on board, of the essential maneuvers to be performed, with which I was more or less familiar. And then another thought would suddenly take hold of me, invade me without my being aware of it, and I would drift at the whim of the winds and currents, letting go of the tiller and forgetting all else. It was as if discontinuity had given way to continuity, to my permanent preoccupations. Only for a spell, of course, but that spell might last an hour or two. The wind could change and clouds could bear down—people would sometimes shout warnings from the shore, but I didn't even notice. I was imprisoned in my thoughts. I lost all awareness of what I was doing, of the water, the wind, the danger, my cockleshell of a boat."

"You nearly drowned, they say."

"More than once, but not at Potsdam. Elsewhere, out at sea in America."

"You compared thought to light, if I remember rightly."

"Well, what of it?"

"So light is also continuous and discontinuous?"

"Don't say you didn't know!"

"I must have known, but—"

Almost brusquely, he cuts her short. He asks her what they're taught at school these days and claims that "in his day," or least from the 1930s onwards, everyone knew about the dual nature of light, being at once a wave and a particle (continuous as a wave, discontinuous as a particle). It was one of the great triumphs of the human mind to have accepted this contradiction, he says, and it's a shame to have forgotten it.

He tells the girl that he has been guided by light since his childhood, that it has captivated and fascinated him, that light is the greatest mystery in the universe, that it undoubtedly holds the key to its entire secret, that every question proceeds from and reverts to it, that it is both birth and death, knit and purl, problem and solution. He tells her that, through the medium of thought, he has more than once striven to become a kind of celestial body launched into space in company with light.

"It's continuous and discontinuous," he says, subsiding a little. "Yes, like most things. I made my debut with light. A German physicist named Max Planck had discovered that, under certain circumstances, the release of energy is not continuous—contrary to what we believed until then and contrary to what was accepted and taught. It's not a regular flow, not a flux. No, if you examine it closely you might think it had hiccups. The fact is, energy releases itself in very small amounts."

"In quanta?"

"There, at least you've said the word. Planck was an exceptional theorist, and we all owe him a great deal. A very inventive mind and scrupulously honest, he sensed that he had put his finger on a sensational phenomenon. He's said to have confided this discovery to his son in Berlin in 1900, but he didn't see, or didn't want to see, all the consequences that flowed from it. I think he sensed that he was calling the world into question—that he was, in his laboratory, drawing near to the ultimate mystery and, at the same time, to potential enlightenment. But he hesitated on the threshold."

"And you crossed it?"

"'Who can tell? I certainly tried to. There were several doors we could have pushed open, and it really seems I chose the right one. Or one of the right ones. So I was told, anyway. And so I'm still told."

"Why did you opt for that particular door? What was special about it?"

Before answering his youthful visitor, he asks if she's studied any math.

"Only a smattering," she replies.

"I ask that question so as to discover what level we should adopt when talking to each other, you understand?

The question of level is of paramount importance. The Dalai Lama came to see me on two occasions. He proved very sensitive to the idea of level. If you ask me a naive question and I give you an erudite reply, you won't understand a thing—you'll get nothing out of it. Neither will I, because I'll be telling you what I already know. On the other hand, if someone asks me a would-be erudite question and I give him a childish reply, both of us will have wasted our time and I'll look ridiculous. My questioner will have made the trip for nothing. So first we must settle on the right level of discussion to adopt."

"What made the Dalai Lama come to see you?"

"I don't know, I never asked him."

"Who gave him your address?"

"He didn't say."

"Did he make an appointment?"

"You'd have to ask Helen that."

"Was it interesting?"

"Yes, once we'd found the correct level. And in his case that wasn't easy because he's very well informed in some respects and very ignorant in others. But on some subjects, like continuity and discontinuity, he caught on very quickly. He has thousands of years of mental agility behind him."

Einstein falls silent. Looking at the girl as if he's seeing her for the first time, he asks, "What about you, by the way? How did you find this address?"

"By a process of elimination."

"Meaning?"

"I looked up all your addresses on the Internet. I made some phone calls, made a few wasted journeys. Often I was shown the door. I got lucky on the eleventh try."

"So why did you come to see me?"

"I have to write something about you."

"An article?"

"Whatever I like."

"You were asked to write about me?"

"It was my own choice. To be honest, I was very eager to see you. The writing business is just a pretext. I knew your name and your face because the whole world knows them. When I was in a bookstore one day, a sales assistant presented me with a bookmark, and this bookmark had a picture of you on it. I kept it. I turned it over and over in my hands for evenings on end, looking at your eyes and wondering who you really were. What were you doing between the pages of my books? Had you really changed my life, as people say? Did you still have something to tell me?"

At this moment the secretary, or at least, the person the girl assumes to be a secretary, the dark, elderly woman named Helen, enters without knocking and deposits a big stack of mail on Einstein's desk.

We see envelopes bearing stamps from different periods, books of recent date, a box of chocolates wrapped in silver foil, newspapers, envelopes adorned with hand-painted flowers, scientific journals from various countries. Some of the latter, when seen in close-up, turn out to be printed in Hebrew and Hindi.

"You still get letters?" the girl asks.

"Yes, as you see. Here and at other addresses."

"You have other addresses?"

"They forward my mail. I can't leave this place."

"Do you reply?"

"I've always endeavored to answer all the questions I'm asked, however preposterous. If only you knew how many cranks explain to me that I was badly mistaken, that the universe was really created by archangels doggedly beating a béchamel sauce, or by a giant composed of luminous dust who has gradually disintegrated but will one day reconstitute himself with the most disastrous consequences from our point of view . . ."

"And you reply to them? All of them?"

"Yes. At least a line or two."

"What do you tell them?"

"I nearly always say they're right and thank them for their help. I tell them that, thanks to them, I'm going to revise my ideas on such and such a subject. They usually leave me in peace after two or three exchanges. If I entered into the least debate they'd keep replying, and our correspondence would never end."

"The questions they ask you—what are they about, generally?"

"They often ask if I've encountered God in curved space-time. That's question number one. They want to know what he told me and in what language. Unless they think I'm Satan, which interests them just as much. A Satan who has decided to make off with science and do away with

God for good. Why is God letting me do this? Why hasn't God struck me down yet? That puzzles them. In his place they wouldn't think twice. They would have taken care to annihilate me for safety's sake. They also request information about the nations living on the sun, or on Jupiter or elsewhere. They suspect me of knowing about them even if, for reasons connected with the arcana of international politics, I stubbornly refuse to divulge those facts. Because I am, of course, involved in a number of conspiracies. They tell me about their visions and the apparitions that have visited them. They quite often refer to Purgatory, which most of them locate in the constellation of Orion, whose residents, some of them celebrities, are acquaintances of theirs. An American footwear merchant sent me a short letter offering to present me with a philosophical solution to all our problems, an answer to all the questions humanity has been asking itself from the very first. He said he would need only fifteen minutes to explain it to me."

"Did you agree to see him?"

"No, not that one, but I'm sure I sent him a reply. I've forgotten. My memory goes misty sometimes, and I lose track of my life. Once, in the United States, I replied to a man on death row who wanted to study physics and asked me how to go about it, having very little time left. On another occasion, at his mother's insistence, I saw a young man who believed himself to be Christ in person. He'd withdrawn into the mountains and refused to come down. We went walking in the woods together for an hour or two.

I tried to point out that Jesus had come down out of the mountains to talk to mankind."

"Did he listen to you?"

"No, no, he was genuinely insane—incapable of listening to anyone. But his mother was counting on me, I don't know why. I'm just a physicist, after all, yet people address me as if I'm a druid, a prophet, a guru, a messiah."

"A false prophet?"

"Exactly. Most of them are extremely critical of me. They're convinced of the existence, somewhere in the bowels of the earth, of an immense force bent on perverting human knowledge—a diabolical force installed there by who knows what or whom. They suspect that Galileo himself, although he sincerely believed in God, was sent to earth by hell. That's because they're sometimes unsure whether the earth is really round and rotating, and they ask me to confirm this from my present whereabouts."

"Even now?"

"Yes, now, four centuries after Galileo! And when I tell them that our great predecessor was right, they suspect me in my turn of being an agent of the kingdom of sin—of being Satan, as I told you. I'm well aware that it's incredible and laughable, but that's the way we're made. Thought is a slow and fragile process. It isn't supreme. It isn't the most widely shared thing in the world, not by a long shot. To a host of timid souls, to know is to be mistaken, to be lost. There's something fey, something magical, within us. We need sorcerers who'll pipe us an everlasting tune and line their own pockets in the process. We prefer belief to knowledge and claptrap to certainty. That's the way it is."

"Your present circumstances could give rise to confusion, you must admit."

"What do you mean?"

"Seeing you here, talking and laughing, one might be forgiven for thinking you come from another world."

Einstein laughs again, though not for as long as the first time. He spreads his arms and folds them, then turns almost serious.

"There is no other world," he says. "Surely you can understand that? There's no world but this one."

"You've taken refuge in a rather unfrequented dimension, let's say."

"Yes, that's possible. It's also what I tell myself when I think about it."

"And that's all you know?"

"About my present situation? Yes, I know nothing, neither the why nor the how. I simply accept it, I can't do otherwise. I don't see how I could get out of here. In order to go where, anyway? To do what?"

The girl remains silent for a few seconds (she temporarily decides not to pursue the subject of another dimension but will return to it). "My questions," she says, "will be very simple."

"Always beware of that word. Questions are never simple, young lady. In fact, it's when they appear simple that they're sometimes at their most complicated."

Einstein goes and sits down behind his desk. His footsteps make no sound on the floorboards, old though they are.

"You must constantly guard against simplicity," he mutters as he goes, "especially outward simplicity. I told you that already. Even so . . ."

"What?"

"It's very tempting, I admit, in fact it's my major temptation, my real desire. To impose a simple perspective on a semblance of chaos . . ."

He doesn't complete the sentence. His drowsy eyes stare into space for a moment or two. Every time he speaks his bushy white mustache twitches in time to his words. The girl wonders aloud why some men hide the area between their nostrils and upper lip. Why do they conceal that part of their face beneath a mustache at the risk of smearing it with food when they eat or finding it awkward to blow their noses? Is it a mask superimposed on the words they utter? A filter? A mere adornment? A product of fashion? A token of virility?

Take Albert Einstein's mustache, for example. Has it assisted him in his research? When he touches himself— when he strokes the hairs of his mustache from time to time—does that gesture trigger a flash of inspiration, an idea, the luminous birth of an equation?

She has spoken normally, but it's as if the air has gone on strike against her voice—as if it hasn't even carried her words. Truth be told, her banal reflections on the masculine mustache in general and Einstein's in particular were not addressed to anyone. They were just a brief, pointless digression, a voice test. Even so, she finds it odd to hear no echo. She feels she hasn't said anything at all.

Seated behind his desk with both hands at rest in front of him, Einstein looks at the girl, smiling again like someone who hasn't heard a word of what she said and, consequently, owes her no reply.

She still gets that periodic feeling of misalignment and disjointedness.

All he says is, "Well, I'm listening."

TWO

SHE OPENS HER BAG and takes out a tape recorder.

"Do you mind?"

"I felt sure you'd already turned something on," he says.

"Certainly not without asking you first."

"You wouldn't believe how often I've been spied on. Spied on and burgled as well, I'm pretty sure, in all the countries I've lived in. Sometimes, in the mornings, I would find my books and papers in a strange state of disarray. People came to pry into my ideas. On the other hand, I used to tell myself that my ideas wouldn't mean a thing to them, so perhaps it was the wind or a neighbor's cat."

"Do you still get these nocturnal visits?"

"No, not here. I don't think so. At least I'm spared that."

"So I can record you?"

"If you like, but be warned: people sometimes record me and end up with nothing on the tape."

"Why?"

"I don't know. The electromagnetic field that surrounds

us, thanks to which science exists and we're able to express ourselves, its finest fruit being my beloved light, plays unexpected tricks on us at times. Don't go imagining that I can explain everything, whatever you do, but in your place I'd also take some notes to be on the safe side."

"Very well."

The girl calmly goes over to the great man's desk and picks up a notepad and an old fountain pen, which she shakes to release the ink. The floorboards creak faintly beneath her feet. She pushes aside a small pipe rack, pulls up a chair and is about to sit down in her turn when she notices a framed engraving on the wall.

She looks twice at the engraving (a black-and-white portrait of a man from days gone by) and takes a few steps towards it.

"Newton?" she asks.

"Yes," says Einstein, "Isaac Newton. You recognized him? Congratulations."

"It wasn't difficult, his name is there at the bottom. Besides, I've seen quite a few photos of the places where you've worked. They often showed this portrait hanging on the wall."

"True. It has accompanied me everywhere."

She takes a closer look at the engraving, then gestures at the door she came in by.

"Wasn't it him I saw with the others in your waiting room?"

Einstein gives a start. "You mean he's still there?" He looks faintly perturbed all of a sudden.

"That was my impression."

Einstein seems not only perturbed but put out, almost exasperated, by what the girl has just told him.

"He's already turned up several times since I've been here," he says, lowering his voice. "He persists in doing so. Having been unable to meet me during my lifetime, he's making the most of the present situation. I find it very hard to get rid of him."

"What does he want?"

"He insists on talking to me, on convincing me."

"Convincing you of your mistakes?"

"Certainly. But I have a bit of trouble with him, mainly because of his temperament. On the one hand, there's no doubt that he's a very great man. One of the greatest. Intuitive and methodical. A giant. He laid down some truly magnificent laws, he justified and explained gravitation, and his calculations are impressive. Even today."

"But?"

"'Well, how can I put it? He held that every cause had an effect and every effect derived from the same cause. It all worked. He thought he'd straightened out the universe, discounting a few minor details. There, he told himself, the universe is as I say it is. I form no hypotheses, I prove everything I put forward, I've fathomed God's innermost thoughts."

"Was he wrong?"

"No, it can't be said that he was wrong. His system held up. From his own point of view he was never wrong. But he saw only one little corner of the world from a certain angle, and in his day he couldn't do otherwise. There were dimensions that escaped him. Those that separate the stars, for instance, or the minuscule dust of matter, atoms and particles. Elements of calculus, instruments. Besides, he was a man from another age. Thought goes with the age and even with the objects that age engenders."

"What do you mean?"

"Where scientific innovations are concerned, or at least the ones Newton should know about, I find it impossible to explain certain things to him, try as I may. I don't know how to talk to him and I waste an awful lot of time. He doesn't know what an elevator is, for example, or an airplane."

"Why an elevator?"

"Come and see."

Einstein gets up and opens one of the doors. Laying aside her notepad and pen, the girl rises too and follows him. She has scarcely taken two steps when she changes her mind and goes back to get her tape recorder, which she hangs on her shoulder by the strap. She hasn't found it too hard to accept the illustrious English physicist's presence in the waiting room. If she has any questions about it—and she must have—she'll ask them later. For instance: How did I get in here? What exactly is happening to me? I feel I'm awake—no way is this a dream. In that case, what? Am I still alive, or is this ill-defined study just an antechamber leading to the unknown shores of the Great Divide?

She recalls old stories, even films, in which incautious travelers unwittingly make the transition from one shore to the other and continue to behave exactly like the living. Meanwhile, the genuine living keep on talking right beside them without even noticing their presence. This surprises the travelers and sometimes annoys them, but there's nothing to be done: sooner or later they have to accept that they're trapped—that the way back is barred to them forever.

Can that be the stage she's reached?

From time to time she rests a hand on the left side of her chest and is reassured to feel her heart still beating. She's also breathing, talking and walking. The floorboards make a sound beneath her feet. She no longer doubts that she's on the right side of the divide. Her main problem, actually, is not knowing how much time she has at her disposal. She doesn't even know if she's still governed by time as in ordinary life. Her watch remains immobilized. What if the fifth door suddenly opens to reveal the stern and forbidding figure of Helen, who announces that her visit, which has only just begun, is already at an end and she must leave?

How much time does she have left? What time? Who is granting it to her? More questions, but they'll keep.

Beyond the open door we find ourselves on a landing immediately opposite an elevator in a stairwell. Einstein points to it.

"You see this elevator?" he says.

"Yes, but what floor are we on?" the girl asks, looking up, down and sideways. "I didn't think I was so high up."

"Don't worry about floors or dimensions and all that stuff. We're precisely where we want to be, or rather, where

we think we are. Surely you must be starting to understand?"

"Yes, but . . ."

She's probably tempted to ask what she ought to be
starting to understand, but she thinks better of it.

She leans over the stairwell. It looks immensely deep,
as if they were at the top of some huge skyscraper. We lean
over too but can't see the bottom, which is bathed in
shadow.

"Let's imagine," Einstein says, "that we're both in that
elevator and the cables suddenly snap."

"We'd fall."

"Yes, we'd fall, us and the elevator together. No, no,
don't be frightened, I'm not going to drag you into that
cage. This is just an experiment in thought. Let's imagine
we're in that elevator, but without any external reference
points, without a sound, without the slightest jolt or vibra-
tion. Are you there?"

She concentrates. He tells her again that she must make
an effort to imagine a silent elevator, a perfect lift, a dream
elevator. She half closes her eyes and breathes evenly.

She's now, in her imagination, on board a remarkably
stable and silent elevator. Is it moving? She can't tell.

"Are you there?" Einstein repeats.

"More or less."

"How would we know we're falling?"

"We'd know when we reached the ground."

"Imagine that there isn't any ground. Imagine that the
elevator doesn't stop. Well?"

Shutting her eyes once more, she tries to imagine an endless, vertiginous descent, a virtual plunge into a virtual abyss. It's a mental exercise that doesn't seem beyond her capabilities.

To help her, he says again, "How would we know we're falling? That we're being attracted downwards? We'd be in a state of weightlessness. For us the world would be abolished—nonexistent, as it were. If we were shut up in our opaque cabin on the ground, on the other hand, our legs wouldn't be able to tell us if we were at rest on the surface of the earth or accelerating upwards into empty space."

She suddenly opens her eyes.

"Gravity is relative, you mean?"

"Yes, good, so you've already grasped that. Motion devoid of disruptions is imperceptible, and everything follows from that. We're imperceptibly revolving around the sun. The solar system is also on the move—as, beyond doubt, is the Milky Way, our galaxy. Even, perhaps, the entire universe."

"You've finally accepted that?"

"That the universe is expanding? Yes. I thought it was static at first, it's true. I shared the general, traditional view and asserted that the star systems moved within a fixed universe, but by degrees all astronomers came down on the side of expansion. I surrendered to the evidence. I'm not pig-headed."

"So everything is moving and we feel nothing?"

"Absolutely nothing. We stretch out on a grassy bank beside a stream, we listen to water trickling between rocks, we feel the breeze stirring the branches of the trees, we see clouds drifting overhead. And that's all. No other sense of motion. We believe we're absolutely immobile on this earth."

"Has it been proved?"

"That we feel nothing?"

"No, that we're moving."

"It was proved a very long time ago, you can take my word for that. Have you ever flown?"

"Yes, of course."

"Come with me."

Einstein goes back into his study and closes the door to the landing and the elevator. Picking up a model airplane the girl hasn't noticed before, he holds it up in one hand like a child playing.

"In a plane," he says, "a big passenger plane, when all is still and we're gliding through the air without any turbulence, without so much as a tremor, there's no indication that we're moving. All the objects around us are moving too, so we feel we aren't moving at all. Like now, on the ground."

"In a railway station," the girl says, "when two trains are standing side by side and one of them pulls out smoothly and silently, it's hard to tell which train is moving, our own or the one next door."

"Exactly."

"The other day, when I was on my way to see you, a boyfriend drove me to the airport. A plane was coming in to land just as we were driving along the highway. Looking at it, I had the distinct impression that it was hovering motionless overhead."

"Because you yourself were moving, and in the same direction."

"And Newton can't grasp that?"

"Yes, of course he can. Galileo already knew it in his day—'Motion is like nothing,' he said. These are essential but elementary lines of reasoning. They've been known for a long time, one way or another, even in default of elevators and airplanes. We move only in relation to something else. Even though trains didn't exist—even though a horse-drawn carriage that didn't make your teeth rattle was inconceivable—certain minds were capable of imagining ideal motion. What was and still is harder is to change scale—to apply everyday observations to the universe as a whole. We see on a small scale, so we think on a small scale, a very small scale, and in straight lines. Our thinking has been shaped by the geometry of the ancient world. For two thousand years we saw the world through Greek eyes—a difficult habit to shake off. Come, I want to show you something else."

He walks across the study with a touch of excitement, followed by the girl, who doesn't let go of her tape recorder. She endeavors to scribble some notes from time to time, but it's very difficult.

"As a scientist," Einstein tells her, "your first obligation is to forget what you see, or rather, what you think you see."

"Like the sun revolving around the earth?"

"Yes. We still say 'the sun is rising' and 'the sun is setting' although we know perfectly well we're wrong to do so. Indeed, we've known it for a long time. Every traditional school of thought tells us that we don't see the world as it really is, that we must liberate our gaze from our habits, but we're caught in the trap of our senses and, consequently, of our mode of speech. Words stick to the tongue. We can't speak without them, yet they lie to us continuously. Here, look at this."

He takes a few inaudible steps and opens another door. From one moment to the next, the two of them find themselves outside on a fine, sunny day. In front of them is a stretch of water with a miniature sailboat gliding gracefully across it. We could believe we're in a municipal park. All that's missing are some children.

The girl seems to have decided not to be surprised anymore, at least for the time being. She might be in some science museum, being shown around by a friendly old guide with long white hair—one who might almost be a phantom himself, a state-of-the-art hologram. But she seems to have decided once again, at least for the present, to preserve her freedom of action, to hold off and refrain from asking questions. And, since we're following her, we do likewise.

"Look at that boat," Einstein says. "Imagine you're on board and it's moving fast. You climb to the top of the mast and drop a tennis ball down its length. Are you with me?"

"I think so."

"Where does the ball land?"

"At the foot of the mast."

Her reply is prompt and instinctive.

"Are you sure? The boat goes on moving while the ball is falling through the air. Did you take that into account?"

The girl reflects for a moment. Then, "Yes, it definitely lands at the foot of the mast."

"Bravo, you're right. You're quite correct. That was a test, did you realize? Yes, yes, I promise you: it's a test that divides people into two schools of thought, even today. That said, I can reassure you: you have a modern cast of mind. Which, just between ourselves, is not surprising."

"Doesn't everyone reply as I did?"

"Oh, no! Some people think it over and make calculations, say this or that. And half the time they're wrong."

"Why does the ball land at the foot of the mast?"

"Because in this context the ball and the boat are one. They belong to the same system. We can't separate one from the other. The same thing applies on board a plane. Toss a coin into the air and it'll fall back into your hand, however high you throw it. Nevertheless, during the second the coin is in the air, the plane will have flown on at full speed. But the coin has moved with it. They're one, they share the same air."

"That's true," she says.

"Yes, it's true, as you say. The two objects, the coin and the plane, seem to us to be separated in the air, but they're one. We could even go a little further. Look over there."

He points to the toy sailboat on the surface of the pond. "Imagine you're on board and watching the ball from the deck. It describes a straight line as it falls."

She looks at the sailboat being propelled across the water by the wind. Seated at the foot of the mast in her mind's eye, she pictures a ball falling.

"Yes," she says. "A straight line. Of course."

"Now imagine you're watching from here on the bank."

"All right."

"The boat is moving and you aren't. You're stationary."

"I'm stationary."

"Seen from the bank, will the ball still appear to be falling in a straight line?"

Under Einstein's amused gaze, the girl debates this question for a moment. She watches the graceful little boat heeling over before the breeze. The answer, she senses, isn't easy. She hesitates. To help her he takes her by the arm and escorts her back into the study.

"Come, I'll show you the whole thing on the blackboard. Blackboards are designed to clarify things."

He leads her over to the blackboard. Drawn on it in chalk—they weren't there before, the girl could swear it—are two silhouettes of the same boat with the ball's trajectory represented by a dotted line, one seen from the shore, the other from the boat itself.

And the two trajectories differ.

"You see?" he says. "Seen from the boat the ball describes a straight line. No doubt about it, everyone agrees on that. Seen from the bank, given that the boat has sailed on, the ball unquestionably describes a curve. Two observers, one on board the boat, the other on the bank, will come to different conclusions."

"And who is right?"

"They both are."

"So it all depends on one's point of view?"

"Of course. Everything depends on the observer's location. As it falls, the ball describes a straight line *and* a curve. It's impossible to choose between the two trajectories. You could make the same observation in a railway station with one observer in a train and another on the platform. And both of them, although they aren't seeing the same thing, would be right. Do you like basketball?"

She's not too sure how to answer. He takes her over to another of the doors and opens it. Instantly, the study is filled with clamorous American voices. In a vast sports hall, a motley, noisy crowd of spectators is watching a contest between two teams of giants, most of them black and so tall that they might be mistaken for creatures from another planet. Agile and dexterous, they're running and jumping around with a big leather ball.

"Look closely at that player over there," Einstein says. "I bet he's going to do what's known as a 'coast to coast.' He'll run from one end of the court to the other, bouncing the ball as he goes. That's the rule of the game—you have to keep bouncing the ball with your hand. But look closely and try to put yourself in his place. To him the ball naturally appears to travel straight ahead as it rebounds from his hand to the ground and vice versa."

"Naturally."

"But to us, who are watching him traverse the court from here, the ball is moving, traversing the court with him, so it describes a broken line. To us, the trajectory of that ball is never straight."

"And we're both right?"

"As in the case of the boat. Imagine the ticking of an alarm clock beside your bed. It's regular, like your heartbeat. Now put the alarm clock beside someone on board an ultra-fast spaceship. Moving at the same speed as his alarm clock, he hears the same regular ticking. But to you, given that you could hear it from so far away, it would sound distorted. It would seem slower to you than to him, just as the basketball's trajectory seems to you to be a broken line, but longer."

"So time expands?"

"Yes indeed. It's called the relativist expansion of time—an inelegant term, I admit, but the thing itself is beautiful!"

They linger in Madison Square Garden for a few moments, watching the ball rather than the game as a whole. There's a deafening blare of trumpets. One of the teams, which is going through a bad patch, has called for time-out. Some booted, high-stepping drum majorettes with crimson lips and dazzling white teeth file out onto the basketball court. There to promote the belief that both teams are going to win, they kick up their bare legs and intone a victory chant.

Einstein turns to the girl and suggests that they go

back. She agrees with a final glance at the gigantic men, who are mopping their faces with white towels.

THE SOUNDS OF NEW YORK fade very quickly when the door closes, like a smothered voice. Einstein resumes speaking as if he has never stopped.

"The dimensions of an object—those of a train door, even of a passenger—vary in certain cases. It all depends on our viewpoint and, possibly, on the speed at which we're moving. So you see what we're obliged to accept, young lady—we, the apostles of rigor: the existence of seemingly antithetical phenomena. You see how hard it is for a scientist to arrive at precise and incontrovertible conclusions. The same goes for light, to which I shall revert for a moment. It used to be pictured as a wave transmitted by some strange and paradoxical substratum, at once rigid and yielding, which was called 'the ether' and enveloped the world. This was based on good, sound reasoning. I did away with the ether. We then discovered that light was composed of minute particles, which were very soon christened photons. We also had our proofs in that respect. So the newcomers of whom I was one—those who asserted that light is corpuscular—were right. Those who described it as undulatory were also right. Incidentally, have you heard of Tycho Brahe?"

No, she hasn't heard of Tycho Brahe. She doesn't even know who or what he's talking about. In a few short sentences, Einstein explains that Tycho Brahe was a man, a Danish astronomer, who lived in the sixteenth century and died in the seventeenth, a superlative observer of the heavens, who succeeded in producing a meticulous account of the motion of the visible stars, drew up celestial maps and astronomical tables that remained in use for a long time—and proved that the earth stands still in the sky.

"How do you mean, he proved it?"

"He thought he'd proved it, and in a very ingenious manner. He took a cannon and fired some cannonballs in an easterly and westerly direction. Cannonballs of the same weight, of course, and propelled by exactly the same charge of gunpowder. If, as Copernicus had claimed, the earth was in constant motion around the sun, the cannonballs fired in a westerly direction would be bound to carry less far than those fired towards the east."

"That was logical."

"Yes, logical from his point of view. If the earth had moved while the projectile was in the air, a cannonball fired towards the west would be bound to have traveled further."

"And it didn't happen?"

"No, not at all. The projectiles fired towards the east and those fired towards the west landed exactly the same distance from the cannon. Conclusion: the earth was well and truly motionless. There had to be another explanation for various phenomena, for instance the alternation of day and night."

"If I've understood you correctly," the girl says, "experiments are not enough."

"They still have to be carefully prepared, they mustn't comply with preconceived ideas, and the conditions under which they're conducted—the conditions are extremely important—must be constantly checked."

"Did people believe this Dane of yours?"

"Yes, for some years. Then along came Galileo. He demonstrated by means of other experiments that the cannon, the cannonballs and the earth were inseparable. It's the same with the boat and the plane. If you want some terrestrial object to gain its independence from our planet, you have to travel far out into space. That was impossible in their day. Inconceivable, even."

"So when we look at the sun it's like being in a railway station? We think we aren't moving—that another train is pulling out?"

"Added to which—not to look very far afield—in the case of our planets, for example, there are several trains moving at the same time. And the tracks aren't the same distance apart."

"Is the station moving too?"

Einstein considers this for a moment or two.

"Yes, even the station," he replies with a smile. "Most certainly. And even the town around the station. And even . . . But that doesn't matter. Forget about trains and planes. Come with me."

Taking her hand once more—she hardly feels his touch—he leads her across the study and back to the third door.

HE OPENS THE THIRD DOOR—the one that led to Madison Square Garden—and they're suddenly in the midst of a starlit sky at night, surrounded by myriads of heavenly bodies.

"Look at the universe," Einstein says. "Or as much of it as you can see, at least. Look . . ."

The night sky is astonishingly clear. It's as if the urban glare that now blots out the stars has been extinguished—as if the fumes we constantly emit have dispersed and the air is as pure and transparent as it doubtless was in days gone by.

As usual on these occasions, they stand there in silence for a while.

"I'm no astronomer," Einstein says at length. "Don't expect me to name all those constellations one by one. There are specialists for that. They're like genealogists with all the celestial families at their fingertips, families that are known as constellations on a minute scale but may also be vast republics of stars—galaxies, in fact. Or systems."

"I've got an uncle like that," the girl says. "In the summer he goes hiking in the hills with a little telescope. He forgets all about his wife and children."

"One can understand that. Look, the universe is irresistible. It's seduction itself. It's beautiful, if that word still has any meaning in those measureless expanses. It embodies all the beauty there is. The sight of it is a cruel, permanent reminder of how diminutive we are. It crushes us. At the same time, it amplifies us by making us welcome. It opens our eyes and, even more so, our minds. It accepts us. The ancients used to say that it reveals itself, manifests itself to us. Unlike God, who's so careful to hide himself away, it's a gigantic exhibitionist."

"Why do you say it welcomes and accepts us? On the contrary, it rejects and marginalizes us. It excludes us."

"What makes you say that?"

"Just about all the things I've read," the girl replies with a faint sigh. "And it's scientists like you who write them. They say we'll never escape from our tiny habitat. We're confined to our little earth forevermore."

"Do you regret that?"

"Sometimes, yes, a bit. The earth has shrunk. We know every part of it. To people my age, that can be disheartening. We've seen everything on earth. What else is there? Setting foot on Mars at vast expense, perhaps, or on some satellite or other. But that's all. As for going beyond the solar system, we can't even dream of it."

"All the same, we went to the moon."

"'Yes," she says, "long before I was born. To me, that's nothing new."

"Do you realize what it represents, going to the moon?"

"What do you mean?"

"All that effort, all that money, all those risks—in terms of the universe, it's a light-second."

"As little as that?"

"Yes. Light covers that distance in a second. And they call it 'space travel'! I ask you, what's the use of going there in the flesh? Thinking is quite sufficient!"

"Thinking of what?"

"Of everything. Listen, young lady, I'm being frank with you. I may not have devoted enough thought to these matters. Listen and you'll help me to think, I assure you. I think better when eyes like yours are looking at me—and when I'm talking to them. The universe that used to be ours is no more, we've got to accept that. For hundreds and thousands of years, at least in the West, we thought of it in dimensions that were almost human, or certainly conceivable and traversable. It was as if we didn't want to let it recede and escape us. It was a universe within eyeshot, almost within reach."

Einstein has suddenly lowered his voice as if the world around them is listening and the time has come to confide a secret. There used, he says, to be messengers commuting between the universe and ourselves, between the sky and the human race. The Greeks had Iris, who used to slide down her rainbow, the Christians had their angels and the Indians their Apsaras. The sky was our garden. We had also established other links with the stars, altogether imaginary links that corresponded to our desires and fears to such an extent that certain configurations of heavenly bodies were believed to influence our destiny and character. In other words, what is still called astrology.

"Look, young lady," he says, brushing aside the universe with a fingertip. "We aspired to read the night sky like a great cosmic book, the work of a single author in which our puny lives would, in one way or another, be recorded. We didn't want to be separated from it—to be banished from the sky. We wanted to listen to what it had to tell us, which couldn't be other than true.

"This gave rise to a strange feeling," he adds, torn once more between two stubbornly conflicting impressions. "The night, which your uncle sees as a carpet of lights and the most marvelous thing life has to offer, was then an enigmatic book of which certain people—as always happens—proclaimed themselves to be the licensed interpreters. They transformed this vault of truth into a simulacrum, an illusion."

"Being distrustful of those who persistently deceived us, we developed the habit of also distrusting the night, which enshrouded us, or so we thought, in gloom and illusion. We put our faith in light alone. We felt unable to give our allegiance to the universe the night delineated for us on a black background. We were illuminated by daylight alone, by the diurnal sun, which certainly cut us down to size but chased away the ever-menacing shadows. Like our intelligence, which dutifully keeps ignorance at bay, our royal master the sun made everything reappear each morning."

"And also made the trees grow," the girl says.

"Yes, as long as it rained. From that point of view nothing has changed," Einstein goes on, "even if we're dispensing with it more and more. The sun of old used to shower us with its light and its blessings—one of which, don't forget, is shade. Without the sun, which was the supreme good, we didn't exist. We subsisted on it. It was, in the true sense of the phrase, our *raison d'être*."

"You still haven't answered my question," says the girl.

"What question?"

"Why did you say that the universe makes us welcome?"

"Ah yes, I was forgetting. You see what happens when I talk about discontinuity? But I'm coming to that. Let's see now. Do you have some idea of what happened in the course of my life?"

"Plenty of things," she replies, rather taken aback. "World wars, technological innovations, revolutions, bloodbaths."

"No, that's not at all what I mean. Every period has known war and carnage. What we witnessed in our time— the number-one phenomenon from every point of view— was that the universe expanded under our gaze to an unimaginable extent. It performed a prodigious leap of as much as fourteen billion light-years. An inconceivable distance. Even one light-year surpasses our sensory and cerebral capacities, so fourteen billion, just imagine . . ."

"But isn't it the other way around? Shouldn't such magnitudes be propelling us in the direction of egocentricity and egoism? Shouldn't we be turning away from such a monster and reinforcing our sense of solitude? Shutting ourselves away instead of opening up? Reverting to our petty quarrels, our minor troubles and cares?"

"Yes, you're right. And for many people, even scientists, that was the initial reaction: infinite discouragement on the same scale as that vast no-man's-land. Confronted by fourteen billion light-years, all we could do was shrug our shoulders. Yet I still maintain that the universe drew nearer to us. And, at the same time, we accepted the night. We tamed it and loved it. We finally learned to read it."

"But how did the universe draw nearer to us?"

"Oh, that's extremely simple for once. It wasn't long before we realized that we're composed of the same matter as the stars."

"Really?"

"Yes, so the astrophysicists assure us. They actually guarantee it. All the shapes you see around us, whether solid or gaseous, are composed of the same basic atoms, the same particles, as ourselves. That matter is the same everywhere: it's the stuff that constitutes you, your skin, your hair, your leather bag, your tape recorder, the air that separates us and enables us to hear each other, the walls of this building, the trams going by in the street, the atmosphere, the stars we see shining at this moment. We call it nuclear because of the atomic nuclei of which it's composed."

"Are there other kinds of matter?"

"So it would seem, but this is the one that interests us: ours."

While listening, the girl can't help glancing at the body of the man beside her. In his brief enumeration, Einstein didn't refer to the matter of which he himself is composed—he took good care not to. So what matter does *he* consist of? And this study, and the model airplane that appeared for only a moment? And that sailboat on the pond?

They stand there without moving, enveloped by the sky. He goes on talking and she goes on listening.

"This universal identity of matter, whatever its location, is profoundly mysterious. It would have dumbfounded the people of the ancient world or the Middle Ages. They'd have found it impossible to believe in it for a moment. Today, who knows? On the one hand, the visible universe is eluding us, becoming unattainable, surpassing our comprehension, defying us; on the other, it has suddenly become familiar. It resembles ourselves in the most intimate way."

"We're made like the stars, you mean?"

"We're made *of* them and by them, and we could, perhaps, study them in ourselves. Look into the depths of the sky. Look calmly at that darkness between the stars and tell yourself that it's populated by other multitudes of stars which we will never perceive. We're millions and billions of light-years apart, yet our protons are absolutely alike. We've lost all sense of proportion in our relationship with the universe. We're off the scale, we're nothing. But one could also say this: we've ceased to be the measure of the world, we *are* the world. What do you think of that?"

"Honestly?"

"Yes."

"I don't find it very welcoming, as you put it. At the level of my protons, my electrons, I'm insensitive."

"That's the problem. That's just it. The universe is everywhere and we can't sense it anywhere."

"Then it's just an idea."

"An idea? No. People have said so, but I don't believe it."

"Why not?"

"If it were just an idea, it could only be our idea. And if it were our idea, why should it be so difficult to access?"

"The world is real, then?"

"Yes, without a doubt. That I've never denied. It's even, in a sense, the definition of the world. The world is reality, the whole of reality. To me, at least. Our eye forms a barrier, of course. It flattens the universe in front of us because that's the way it sees things. We're never really *in* the sky. We're on earth, and from here, our dwelling place, our platform, we see the sky. Quite simply, we look at it. But every look is a lie, every look deforms. There's nothing more suspect than a look. Our eye—our brain, in other words, since one won't work without the other—is terribly restrictive of the world and persists in misleading us. All it does is see. It divorces us from reality and leaves us outside the world. That's why we have to go further."

"You say the world is real."

"Yes, I do."

"Even the world we're in tonight?"

Einstein runs his fingers through his hair and mutters (he has said this already) that there's only one world, but it's profound and changeable. It contains representations of the world so faithful that they sometimes take us in. He speaks of mirrors, of changes of angle and *trompe l'œil*. He's mumbling, so the girl can't hear too clearly what he's saying. She probably decides to raise the subject again later on. If she has time.

Confronted by immensity, our fellow creature, they fall silent once more.

He mutters that this unfathomable world contains no single fixed point from which we could survey it. We can't step back—we possess no privileged observatory.

"Do you have any other examples of what you were saying about the eye and vision?" the girl asks. (We still don't know whether she's a student or a journalist.)

"Lots of them. The atom, for one. Around the beginning of the twentieth century, when the atom had become self-evident even before it was dissected into particles, when it seemed impossible to dispense with it for much longer and I was one of the very first to say so, do you know what we were told?"

"No, what?"

"That the atom was invisible even under a microscope, so it didn't exist. The eye inspired confidence. We could see microbes, so microbes existed, and besides, they transmitted diseases. But atoms? No, out of the question. I can't tell you how many times I heard people say, 'We don't need your atoms!' or, 'Show them to us!'"

"That must have annoyed you."

"To put it mildly."

She turns once more to face the cosmos and closes her eyes, scarcely breathing. Perhaps she's trying not to look, not to see—to effect a surreptitious penetration and violation of the very substance of things, which is also her own.

"Be sure to keep telling yourself," Einstein says, "that

all you see, and all that appears to you to be motionless, is in motion, and in all directions. Our planet is revolving even though we don't feel it. The solar system is in motion, our galaxy is moving, all the galaxies are moving. The universe itself, part of which you see in front of you, is expanding . . ."

"So you already said," the girl remarks, opening her eyes again.

"I never tire of doing so. But I'm repeating it for a specific reason."

"Which is?"

"To get you to imagine, for once in your life at least, what a headache this represents from our point of view."

"It hampers all your calculations?"

"Yes, it makes it hard to compute things precisely—to envision the result of a calculation and then verify it. What's more, some of the stars we see are already dead, as you're no doubt aware."

"Dead?"

"You don't know? You haven't heard this?"

"Yes, but pretend I haven't."

The eternal question of level . . . How much does she know, this girl who appears to be studying for a degree (in journalism?) or simply gathering material for an article? Where has she been studying? What stage of knowledge or ignorance has she reached or stopped at?

If Albert Einstein prolonged this interview, he would be within his rights to question her. Would he have to go back to elementary concepts? Could a man like him be expected to start from scratch? To recite the alphabet he himself perfected?

He could even cut the proceedings short. He would be entitled to stop here—to say: You can't understand me, admit it. He told the girl he answered all his letters, however preposterous, but he could very well make an exception in her case. Forgive me, he could say, but I have so much work to do and I'm behind.

This is where his visitor's smile, her touch of impudence and her physical charms may come into play. Those, coupled with the impression that, possibly because of some imperceptible slippage, the great human clock has stopped for once; that by gambling on space-time the girl has won, and that Albert Einstein, in his present whereabouts, "has time" for her. All the valid reasons he might adduce about his work, his backlog of correspondence and interrupted research, would suddenly lose all cogency and significance.

Whether he speaks for an hour more or less won't cost him much or affect his timetable: he doesn't have one. If it does, he won't even realize it. Besides, as he himself has told her, a visit that compels him to retrace his very first steps may help him to advance further into difficult terrain.

"Strictly speaking," he says, "the distances we have to calculate are unimaginable. We speak of a light-year, that's to say, the distance light travels in a year, without the slightest pause or deflection, at a speed of three hundred thousand kilometers per second. Does that mean anything to you?"

"Yes, of course. Everyone knows that."

"Let's hope so. Well, when the light of very distant stars reaches us, sometimes after billions of light-years—"

"Billions?"

"Yes, billions, I already told you. At least try to pay attention. It takes anything up to twelve or fifteen billion light-years to reach us. I wasn't responsible for those computations."

"Twelve billion or fifteen billion amount to more or less the same thing, don't they?"

"You think so?"

"Yes, I do. I don't see any difference. What if you called it twenty-five or even thirty billion? What difference would it make? In any case, it's meaningless. To me, at least."

"You're absolutely right. Whatever you do, don't try to picture those distances in terms of our measly little kilometers. We can't do that. The prime requirement is to discard all our human references. If it weren't for science, those dimensions would surpass our units of measurement and our understanding. We can't get there on foot."

"So how *do* you get there?"

"In another way. We try to, at least. It's up to us to provide ourselves with aids other than the ones we devised long ago: first numerals, then mathematics. To escape the pitfalls of our senses, of representation, of ordinary thought and limited logic. To equip ourselves, when tackling the universe, with other weapons and different terminology. To be able to write fifteen billion light-years on a blackboard, on a sheet of paper, on a fingernail. In the sixteenth century, one persevering gentleman managed to inscribe the entire text of the Bible on the inside of a nutshell. That's the stage we're at."

"Could anyone read it?"

"No, but it was the Bible nonetheless. One could have verified that with a microscope."

The girl takes another look at the scintillating sky.

"So," she says, "where are these dead but still shining stars?"

"All over the place, among the ones we can see. Some of them have been dead for a very long time. All we have available is their original light on its long journey to us."

"You mean their light has been hurtling unsupported through space for all those billions of years?"

"All those billions of light-years."

- THREE -

THEY LINGER THERE in silence for a while—how long the girl can't tell—with their arms at their sides and their eyes, which they close at times, directed at the cosmos.

"Is that what relativity is?" she asks at length. "The fact that what we see isn't necessarily what exists? That everything is in constant motion? That our viewpoints become modified?"

"No, not exactly, but that's a start. That's the initial attitude of mind. It's indispensable, as I told you. First you have to understand—or accept, anyway—that relativity is in opposition to the absolute. The words themselves imply that, don't they?"

"Are absolutes independent of the observer, as you said?"

"Yes, that's one way of putting it. But you mustn't imagine that, when one says 'everything is relative,' it's the whole story. Relativity itself is in search of an absolute, of course, or of several. If not, it wouldn't be a scientific adventure. It looks for fixed points, invariants, constants. That was my life's work, and apparently it still is."

"How do you proceed?"

"Ah, this is where things become very complicated."

"Really complicated?"

"Look."

"What?"

He beckons her over to the blackboard again. The boat drawings have disappeared and the entire surface is covered with a jumble of overlapping equations.

"I see what you mean," says the girl, who's defeated by the whole thing.

"And that's not all," Einstein says. "Look."

All at once, like luminous projections in a high-class disco, equations, calculations and graphs in white cover the walls of the study themselves, and the ceiling, and the floor, and every stick of furniture, and all the objects including the violin and the music stand. It's a special effect, a form of global decoration. We could well believe we've been propelled into the midst of another star-spangled sky, as dazzling as the first but composed exclusively of mathematical constellations.

The girl, who can no longer feel the floor beneath her feet and might be hovering in midair, seems as impressed as she was by the real sky she has just left. She looks as if she doesn't dare move for fear of disturbing the smallest detail of this indoor cosmos. She knows, as we all do, that nobody plays around unscathed with the matter of which the universe is composed.

"One more question," she asks softly.

"Ask away."

"Or even two."

"That comes to the same thing."

"What happened in 1905?"

Albert Einstein's shoulders sag a little, as if age has caught him unawares for the first time since he admitted this visitor. He shuffles over to his desk and slowly subsides onto the chair he was occupying when we first saw him.

The mathematical symbols silently vanish, the illuminated equations are extinguished. The room more or less returns to its original state.

"Oh," Einstein says wearily, "a few youthful discoveries . . ."

"Can you talk about them?"

"It's over a hundred years ago."

"Quite so."

"Three or four brief articles in a physics journal whose editor was keen to publish me. People have written about them so often. Why bring them up again?"

"I tried to read those articles. I failed."

"You aren't the only one. Anyway, don't bother. Scientific terminology has completely changed since then. Even I might have trouble rereading them, and I'm sure I'd be tempted to correct them if I did—to insert question marks in the margins."

"So reading them wouldn't be worth my while?"

"No, I told you. What you've just seen here in this room—all those interminable calculations, all those assumptions, manipulations and verifications, all that—is just a dark cloud through which we have to pass in order to convince our colleagues of the truth of our conclusions by means of our procedure itself. It's the jargon of our club. We have to conform to it or our membership isn't renewed and we're refused admission. No need for you to venture into it, you'd risk getting lost. After all, you don't refer to the original text when you read something translated from Chinese, you trust the translator. Don't bother!"

"Much obliged," she says.

HE WAS TWENTY-SIX at the time. A physicist by training, he had studied at Zurich's Federal Institute of Technology, the *Polytechnicum*, where legend has quite erroneously branded him a mediocre student. For want of any better occupation, he was working as an "engineering expert" at the Swiss patent office in Bern, where he checked submissions with acknowledged competence—even, perhaps, with interest. Together with one or two friends and his first wife, Mileva, who had been admitted to the Institute like himself, he continued to pursue his enthusiasm for physics, publishing the odd article and making a modest living in the interim.

Everyone was going in for physics at that time. It was the fashionable science, the middle-class science with links to industry, the one that was leading the field and boldly tackling the mysteries of the universe. In Europe it exerted a fascination primarily on steel manufacturers and heads of state.

Einstein meditated on the principle of relativity and extended it to all physical phenomena. In particular, that same year, he analyzed the Scottish botanist Robert Brown's old experiments into the movements that bear his name (movements connected with pollen particles suspended in water) and Planck's far more recent experiments with radiant energy, the so-called black-body radiation emitted by a solid body according to its temperature. By clarifying and interpreting these observations, Einstein at a stroke cast doubt on the old, Newtonian organization of the world, the concepts of space and time. He threw open the door to the atom; he likened energy to matter; he established the speed of light as an invariable and unsurpassable parameter in the universe, as an "absolute" (whatever the location and velocity of the light-emitting object); and, finally, he embarked on the trail of "invariants" with the idea of (special) relativity, a theory according to which the laws of physics remain the same for any "inertial" observer, in other words, one traveling in a straight line at a constant speed and in any direction.

Which was a great deal for one man to achieve in a single year.

The girl has noticed Einstein's sagging shoulders. "Why so tired suddenly?" she asks. "Shall I get you a glass of water?"

"No, no, I don't need water or food anymore, as you

can imagine. That's why I haven't offered you any refreshment, incidentally. Please forgive me. If you're thirsty . . ."

"I'm not thirsty," she says. "I simply wanted to know the reasons for your present attitude. You haven't been the same for the last few seconds, not since I mentioned 1905."

"It'll pass," he says.

"What were you thinking about?"

"Sometimes, in spite of my present situation, memories of those days come back to torment or distress me. I'm like everyone else: I can't help it. Bitter but pleasant memories. Memories of my research, of my friends, of our walks in the mountains, where sometimes, halfway through a sentence, ideas would spring to our minds. Memories of our discussions about light in the darkness of night, of my very long nights of inspiration and my struggles. Of my doubts, too, and misunderstandings. Of discouragement and derision. Of the failure of my marriage—my own fault, no doubt—and my children's rather unenviable lot. Memories of all I've failed to do."

Above all, as he tries to explain without spelling it out, he recalls the time when he, who was to bring about an abiding change in everyone's view of the world, suddenly found himself alone and forlorn in a state of universal incomprehension. The ideas he advanced amid the abominations of the first great European war, which was soon to become worldwide, would in less than forty years present humanity with the keys and powers of which it had dreamed from the first, among them the ability to destroy itself for good and all.

At that time, however, no one was capable of discerning or understanding this. No one, not even Einstein himself.

Never in the history of ideas has there been such a contrast between the secret intuition of one man and the lives and destinies of all.

"Could you give me an example?" the girl asks. "Just one example I'm capable of understanding?"

"That's easy enough," he replies without thinking. "You told me when you entered this study that, according to me, time doesn't exist."

"That's right."

"Although I didn't take you up on it then, what you said was inaccurate. Caricaturists credited me with that statement, but I've never actually uttered or written it. Not in that form, which is so oversimplified as to be absurd."

"What *did* you say, then?"

"I said that, to a careful scientist, the forms we ascribe to time—the words 'before,' 'after,' 'at the same moment,' 'soon'—possess no exact meaning. I said that those forms remain vague, that they depend on the situation and sensations of each and are only figures of speech. Time in that sense belongs to everyday conversation, to the drawing room. I said that time, which seemed unassailable, and which we'd always called our lord and master, is not an absolute value that manifests itself in the same way throughout the universe. I said that time itself is relative and subject to events—in short, to velocity and matter. Besides, we can observe that in our daily lives."

"Really?"

"Sit down on that chair."

He points to an ordinary wickerwork chair. She sits down on it. Einstein gets to his feet. He seems to have recovered his spirits—he's even smiling again. Coming out from behind his desk, he walks over to his youthful visitor—still without a sound—and plonks himself down on her lap.

"Don't be alarmed," he says, "I don't weigh much."

It's true, she can scarcely feel him, but she doesn't speak. She hears him say, "Be that as it may, if I sit on your lap for a minute the time will seem very short. That's because you're an attractive girl. The touch of your skin must be delightful."

He promptly gets up again.

"But if I sit down on a hot stove," he goes on, "the same minute will seem interminable."

"That I can understand," she says.

"Anyone could. What I've just done is very simplistic, of course, and entirely subjective. There's nothing scientific about it, but all the same, it's a first step. Change the circumstances—modify your conditions of perception, observation and study, modify the speed of objects—and time won't seem the same to you. It will cease to seem the supreme and immutable master people have tried to turn into a god."

"Does the same apply to space?"

"Of course. Length contracts in the direction of movement. When we watch a car going past, or even a cyclist, the shape we see isn't the same as if we were seeing it at rest. Objects appear to be contracted by motion. Here too, we can begin with things that are extremely simple—childish, even. As we see it, what looks big to an ant looks small to an elephant. Do you agree?"

"Yes, of course."

"To put it more simply still: to the ant the elephant looks big, to the elephant the ant looks small. At all events, that's the way we talk. However, if you look at the ant and the elephant from a very long way off—long relative to the universe, let's say, or just to the solar system—their difference in size will be imperceptible. And it's the same if you launch yourself into the infinitely small, which is even harder to imagine, perhaps, than the infinitely big. In the eyes of a particle, if I may put it that way, the ant and the elephant are of more or less the same volume. They occupy the same space. The ant may be too enormous for the particle to be able to 'see' it, so what about the elephant?"

"Are we still talking about relativity?"

"More or less. We are, in fact, on its periphery—we're circling it. But let's be quite clear: in saying that, we aren't even talking about the relativity the French call *restreint*, restricted or limited, but which I prefer to call 'special.'"

"So what sort of relativity *are* we talking about?"

"Basic relativity. The everyday kind."

"If you say so."

"Yes, I'm well aware that the simplest things can be the hardest to accept. As I already told you, we've been shaped by thousands of years of blinkered thought entirely subordinated to legends and beliefs. It was a mode of thought on our own scale, one that sufficed for us and imagined the universe far more than it observed it. Do you know the first argument that was leveled at Galileo when he repeated the conclusions Copernicus had reached eighty years earlier and told his ecclesiastical judges, with a whole host of reservations, that the earth may not be static in the center of the universe?"

"No. What did they say?"

"That because God had made man in his own image, consecrated him the masterpiece of Creation and placed him on earth, the earth must necessarily be the center of the universe. The other heavenly bodies, beginning with the sun, could not but revolve around the earth in reverence. They were its vassals."

"But who said that God created man in his own image? Other men, wasn't it?"

"Yes, of course. The authors of the Bible and its commentators, the Fathers of the Church in council—'the authorities,' as they were called. That line of reasoning was utterly circular, yet none of the traditional authorities thought it needed revising. Darwin was told the same thing centuries later: that human beings and animals couldn't possibly have a common origin. Reread Genesis, reread the story of Noah's Ark! Theology and so-called sacred history, including works of self-evidently human authorship, arrogantly decreed that a scientific discovery was ill-founded. Insanity, genuine folly! No one would dare to resort to it today, discounting a few archaic individuals."

"There still are some, it seems."

"And will be for a long time yet, let's not cherish any illusions on that score. I've never understood why, but some of our fellow mortals feel perpetually embroiled in a vast shipwreck. They're in a state of chronic distress and cling to the past like a lifeboat in the teeth of a storm. Well, so much the worse for them."

Einstein gives another example of what he calls "basic relativity," the everyday kind. In those days he used to like to say of himself, "If the theory of relativity proves valid, Germany will claim me as a German and France will proclaim that I'm a citizen of the world. If my theory is disproved, France will say I'm a German and Germany will proclaim that I'm a Jew."

So relativity itself is relative. Basically relative. He finds that amusing.

Just as he has spoken of time, so he now speaks of the words we use to situate objects and ourselves in ordinary space.

"The sky and the sun aren't 'above' us. Strictly speaking, that's meaningless. What about the planets Mars and Saturn? Are they above or below? Above or below whom or what? One just can't talk in those terms. Is what is above us above Australia or below it? The whole thing's absurd, a geographical and grammatical archaism. The sky isn't 'above' us because we raise our heads to look at it. No stellar system or galaxy is above or below anything. The universe isn't constructed like a house based on architectural drawings and elevations. Top, bottom, near, far—none of those words possesses other than a relative meaning, if you give the matter two minutes' thought."

"Is that why you spoke of space-time?"

"Yes, in part. So as to coordinate events, to situate them both in space and in time. So as to find a new absolute on which we could base some serious work. To be honest, the idea came from one of my professors, a man named Minkowski. At first, when I took it up and developed it, no one paid it even cursory attention. I had to wait fourteen years for an English astronomer named Arthur Eddington—the Astronomer Royal, no less—to prove by observing the stars that I was right, and that space-time is curved—or, more precisely, that it curves in the vicinity of the sun. To put it another way, I situated matter in space-time, which curved in consequence. It bowed to me, so to speak."

"Isn't time the same everywhere?" the girl asks gravely.

"Time *isn't*. It has no 'being' and no existence. We can't speak of it as we would of a person, a thing, an element, a substance, an occurrence. We can't say 'it isn't the same,' because that might imply that there are several times and we could compare them, or that the attributes of time change according to circumstances. You see how deceptive our vocabulary is in that respect too, and how our way of speaking colors our mode of thought?"

"What *can* we say, then?"

"That time doesn't always flow in the same manner—but can one even say that it 'flows,' which promptly puts one in mind of a river? That it can definitely expand—yes, at least to our eyes. That common notions like past, present and future are only an illusion, no matter how deep-seated. That these everyday ideas are truly 'relative.' That our perception of them is variable, even though we've tried to standardize it, to indicate it precisely with the aid of clocks—even though we can all see the same time on our wrists. But that isn't *time*, of course. It's only an image of time, an indication on a dial, a convenient convention."

"And space varies too. It seems elastic to us. Neither space nor time is an absolute entity, fixed once and for all outside the very existence of the world. Both depend on objects, on events, on matter."

"In that case, if I managed to take a high-speed trip through space, would I age at a different rate?"

"Most definitely."

"Would I be younger when I returned to earth?"

"Younger than your twin brother, who stayed at home. My friend Langevin came up with that amusing idea. He was right, too. The thing is technically unachievable with human beings, but it's physically possible. Experiments have been conducted with atomic clocks, and they work. Positive effects have been observed. When the clocks return from elsewhere, they're slow."

"Where do they do these experiments?"

"On an infinitesimal scale in airplanes, or in those huge machines for cracking the invisible known as particle accelerators."

"What did the Englishman actually see?"

"Eddington? Well, he spent a long time observing the

sky on Principe, a small Portuguese-owned island in the southern hemisphere. He waited for a solar eclipse so as to photograph some stars from there and see if they really were where I'd said they would be."

"And were they?"

"Yes. The rays of light underwent modifications, spatial deflections, as I'd surmised. I said that this would enable us to see a particular star situated on the far side of the sun, where it was hidden from us. And it was visible. Eddington concluded that I was right. By publishing his photographs he made me a celebrity overnight. People quoted my name everywhere without really knowing why. You have no idea the nonsense that was written about me, especially by the 'authorities' whom I was upsetting. They attacked me. They said I was mentally ill, deranged, raving mad, a public menace—that I ought to be arrested and silenced. But we were very soon joined by other physicists, because evidence is more infectious than lies. The truth can't be resisted for long. Ideas can be stubborn—they stagnate for centuries—but at other times they travel at a fantastic rate. It's as if they're hovering asleep in the air and a passerby wakes them."

"What was this new evidence?"

"In the case of the first form of relativity, which I term special, the expansion of time was accepted. By verifying the curvature of rays of light, Eddington corroborated general relativity. I'll try to make it simple for you. In order to understand the relationship between curvature and gravitation, look at this terrestrial globe." She does so. "Now place your finger on one meridian and I'll place mine on another. If we move them northwards"—she moves hers and he moves his—"our fingers draw nearer to each other as if they're genuinely attracted. It's simply an effect of curvature, but we regard it as a force!"

"The atmosphere in which we live, that of the earth, contains air, which curbs the acceleration of bodies and, in consequence, disrupted our calculations for a long time. In order to attain the true terrain of physics you must forget about air and imagine a physics of space. And thought alone can get there. Only thought can dispense with air."

"I did have some pretty vague ideas about these things," the girl says. "I was susceptible to the infection you mentioned—I thought the subject might interest me in spite of the effort involved. I didn't want to die a fool like a lot of people who have already decided to do so, even at my age. That's why I said to myself: Let's go and see him. May I tell you something?"

"Of course."

One rainy day some years earlier, in her grandfather's barn in the country, she had come across a tattered old collection of sci-fi magazines from the 1950s and leafed through them. What had stuck in her mind was a short story about an alien spaceship in serious difficulties, which was transmitting distress signals as it neared the earth. All air traffic was promptly halted. The spaceship was located and tracked, contact established with the crew and an airport made ready to receive them. It was an extraordinary occurrence—nothing like it had ever happened before.

Reporters flocked to the scene. Everyone eagerly scanned the sky, which remained empty. The distress signals could still be heard—they became ever louder and more distinct—but there was nothing to be seen. It was impossible to locate the machine with the naked eye, or even on a radar screen. All at once, when those on earth were expecting it to land, all contact was lost. A strange sound was heard, then nothing. There was no trace of the spaceship. It had vanished into thin air without anyone seeing it, just when the crew had announced that they'd reached the airport and were about to touch down.

Inquiries were launched, experts called in, various theories advanced. Then the key to the mystery was revealed at last: the spacecraft was so minute, smaller than the head of a pin, that it had plunged into a puddle on the runway.

The fantastic little machine from far away had simply sunk, together with its crew and equipment. No one would ever find it.

"That's quite conceivable," Einstein tells the girl. "We never think of the scale of things. We're always minute compared to something."

"Or colossal, gigantic."

"Or illusory."

She regards her host with a hint of suspicion. It's as if, having reverted to the current situation, she suddenly doubts his physical reality. But he merely gives her a broad smile. Even his mustache is smiling. In a quiet voice, he observes that outside the solar system the star nearest to us is a little over four light-years away, or around forty thousand billion kilometers. And that's only in the very next street.

"How near is the sun?"

"Eight minutes away."

"Is everything possible in the realm of experimental thought, as you said?"

"No, not everything. Human thought has its own limits, but it clearly enables us to go further than immediate reality—further than our senses, anyway. We'd be crippled without it. Do you enjoy thinking?"

"Me?"

"It's wonderful exercise. Thinking . . . I don't know what it is, exactly, but it appears to take place in our ten billion neurons by way of a multitude of contacts and convergences that we more or less control, our mental processes being strange. It's supposed to be a matter of networks and connections, like the plot of a novel of espionage. What I *can* tell you is that nothing else has ever given me such joy, such a sense of tranquility and fulfillment. Look, while we're on the subject of time and space, there's something else . . ."

He beckons the girl out into the star-spangled sky again. She follows him without hesitation.

Pointing to the stars that surround them as before, he says, "When we speak of time, we always say that time passes, that we don't have the time, that things do or don't happen at the same time. But what does 'at the same time' mean?"

"Simultaneously?"

"Yes, but please be more precise. Scientists are always expected to be precise. I told you that all heavenly bodies are in motion, and that the signals they emit reach us at the speed of light. But how do I know that out there, very, very far away, it's the same 'moment' as here. Give it a little thought. Think."

The girl preserves a brief silence, trying to understand what she's supposed to feel. Meanwhile, Einstein adds, "Billions of years may have elapsed by the time I receive a particular signal. Well? How can we establish universal simultaneity? It's impossible, as I'm sure you realize. All this light would have to be immediate—transmitted and received instantaneously. And we know that it isn't like that, that the speed of light is limited. We've known that for at least two centuries. In order to write the history of the universe, we have to cut it into slices. Slices of simultaneity."

He pauses, then asks himself, "Does the word 'measure' still possess any meaning?"

"In music, perhaps?" she says.

He nods. "In music, yes, definitely," he murmurs, "but that kind of measure is our own. We decided on it ourselves, and we have a conductor to tell us if we're obeying it or not."

There's another brief silence during which Einstein's visitor continues to muse as she gazes into infinity.

"Shall we go back inside?" he suggests.

EITHER THEN OR A LITTLE LATER, they revert to the subject of thought. This seems close to the girl's heart, because (although she conceals the fact) she has a nodding acquaintance with philosophy. "To think differently" is easy to say, but how to go about it? The species that styles itself human has always been consumed with admiration for its own thought processes and the prodigious achievements of its "mind," which it has been careful to separate from its crude, perishable body. It has regarded thought as proof positive of its superior origin and privileged status, its medal of excellence and the attribute that differentiates it from other life forms. Even the Vedas called it "divine." For Descartes, human existence was guaranteed by that and that alone.

How to forget that, how to put it to one side?

How to find a judge that is not only equitable but acceptable to thought itself?

The girl makes a diffident allusion to Immanuel Kant, Einstein's compatriot, who had stated two centuries earlier that we must put reason on permanent trial before its own tribunal. By that he meant—as Descartes had already said—that we must beware of glib talkers, conjurers and magicians of all kinds, who are quick to sell us solutions to the world's mysteries; conjurers and magicians among whom we ourselves may be numbered without ever realizing it (like Galileo's judges, who mistook themselves for purveyors of the truth).

Kant meant that reason must be brought before its own tribunal without recourse to extraneous authorities or self-proclaimed experts, allegedly divine texts or hackneyed traditions—that it must be put on trial before itself, cold and naked, and treated with severity.

Einstein knows these writings by heart. In the case of physics, however, when faced with inevitable contradictions, with the relativization of time and space, which used to be regarded as immutable reference points but have now been set aside, how are we to proceed and form judgements? Logic and rigor are not enough. Reality curves and flexes, strays and becomes entangled. Human thought is venturing into territory where there are no maps to guide us, no landmarks, no ferrymen, no bivouacs awaiting us, and where there are even undetectable traps that have assumed the name "black holes." It's territory where there's no top or bottom, no ahead or behind, no temporal succession, no past or present; where we walk along without moving forwards—or possibly, even, go backwards.

How are we to think in the absence of these things? Proceeding from what? Making for what? Will those two new absolutes, the speed of light and space-time, be adequate to guide us?

And what to make of the celebrated questions, undeniably naive but nonetheless persistent, about the order and destiny of the world—questions which are forever renewing their attack to such an extent that the girl herself is still sometimes troubled by them. Why is she alive? Why must she die someday? What is the meaning of her life? Of what secret plan is she a part? Why is the Universe—with a capital U—constituted in this way and not in another?

What is being concealed from her?

Einstein tries to explain how easy and commonplace—and dangerous—it is to proceed from the question "How?" which is the one perseveringly posed by all the scientists on our planet, to the question "Why?": to putative finality. Our mind being what it is, and only fleetingly active on earth, most of us demand that everything have an objective, a *raison d'être*. We see everything on our own scale and in our own image. During our terrestrial lifetime we behave in a particular manner because we all, consciously or not, have an objective. We aspire to lead a long and healthy life, to become rich and powerful (exceptions apart, though the vows of chastity and poverty are also a response to a desire), to acquire material possessions that appeal to us, to attract a particular man or woman, to occupy some senior or junior position, to find the answer to one or several questions, and so on. It is in our nature to desire—sometimes, even, to desire to rid ourselves of desire.

Most of us are constituted in such a way that we also aspire to understand things, that is to say, to reduce them to the limits of our mind, which is simply our reason, and also to understand the wherefores of the "Why?" and all that prompts us to try to understand; in other words, thought, its mode of operation, its boundaries, and so on ad infinitum.

And all this, only to be told that the "explanation" of phenomena is not only impossible but devoid of meaning. That thought cannot explain what has been or is being thought. That that which is not being thought eludes it.

That to aspire to understand the incomprehensible is futile.

We seem to have taken this very need for an objective, a design, the first step towards a final answer, and applied it to the universe as if everything had to proceed in a direction comparable to our own. As if galaxies and electrons were copying our procedure and matching all our internal movements. As if sense dominated reality. As if, where the infinitely great and small are concerned, everything complied with someone's will, with a model constructed by a skillfully concealed architect. As if the stars were hung in the sky to answer our long-standing questions.

"Is that why we invented the gods?" the girl asks.

"It's one of the reasons, no doubt. Possibly the main one. Our limited awareness probably needed another pure and universal dimension of consciousness."

"People must often have asked you that question, I suppose."

"Innumerable times. It's a question that plagued me throughout my life. 'Do you believe in God? Do you belong to a religion? Do you pray?' An obsessive, unhealthy form of inquiry indicative of profound uneasiness and doubt. And it's absolutely true that, when confronted by the infinite beauty of the universe and its sublime harmony, I sometimes experience a feeling I've ventured to call religious. Why not?"

"But it doesn't mean you believe in God?"

"Certainly not. In the light of this boundless magnificence, the notion of a divine creator and ferocious chastiser of the only human race strikes me as wholly absurd. Why should such a genius, who encompasses all things in a cosmic dream of unattainable dimensions, pore over our tiny peccadilloes like some persnickety schoolmaster? Besides, as you're doubtless aware, to a scientist all our actions are predetermined, or nearly all. Our free will is extremely limited. In that case, why should any god punish people for acts they couldn't help committing—acts which the god himself devised for them *ab initio?* The gulf between the magnificence of the universe and God's punctilious cruelty is too wide. He's truly on our own scale and born of ourselves. He's unworthy of the universe."

"And the other reasons?"

"The often-cited desire—indeed, the need—for an authority, for someone to turn to, for an omnipotent overlord who will lend us his ear and support. And certain schools of thought go so far as to believe in absolute determinism: we're either saved or doomed as soon as we arrive on earth; the die is already cast whatever we do. That's almost scientific, after all. Added to which there's our universal fear of death, of returning to the void. The need for justice in another world since we're denied it in this one, et cetera. It's all pretty banal. What else is there to say?"

"But the primary reason is the need for answers to our 'whys'?"

So he believes, Einstein replies, but he isn't sure. He hesitates—he doesn't claim to be an expert on philosophy, far from it. He has only an imperfect knowledge of its concepts and lines of reasoning. He feels an affinity to Spinoza's flexibility, even to his intelligent ambiguity, but any logical and didactic system of thought—theology as a whole, for example—makes him uneasy.

Nothing irritates him more than the sporadic "surveys" conducted by some of the periodicals he receives. On "God and Science," for instance. He finds it impossible to discern the smallest relationship or point of comparison between those two words. What he insists on calling an "experiment in thought" seeks other alliances, other speculations, other approaches, other proofs and points of view, not overly hackneyed lines of reasoning. What is also required before unleashing one's imagination is, above all, a bedrock of honest knowledge and a rigorously observed method.

Without imagination our route remains unproductive. Humdrum and repetitive, it leads from one form of ignorance to another. Knowledge connotes invention. If it doesn't, we go plodding along in the ruts of yesteryear. We mark time on the spot, superimpose other words on things we've already named, and think we're advancing through a wood when all we see are the trees.

Einstein says that he has also known since his youth how difficult it is to slough off our thoughts, our mind, our feelings and references. Imperceptible and insignificant denizens of the world's vast shore, consigned to a minuscule grain of sand, remote from all else and ignored by all that is not ourselves, we nonetheless wish with all our might that this endless beach, of which we never visit more than three or four grains of sand near our own, were comprehensible, accessible and submissive. A residue, no doubt, of the stubborn old belief in the supremacy of man, the focal point and magnum opus of Creation, the less than faithful image of a god.

"Even where I am now," he says, "disburdened of the weight of my body and exempt from hunger and thirst, from physical pain, from everyday concerns and even from the desire to smoke—even now that I'm relieved of all human needs and ambitions and have no fear of death, I can't emerge from myself. I'm still obsessed with understanding a world in which I no longer have to live. Even today, believe it or not, I'm absentminded and bump into the furniture. I haven't changed from that point of view. Once I start thinking I abstract myself from a world in which I no longer am."

"Like the old days?" she says.

"Yes, like the old days. Like when I went sailing, or like the day in Los Angeles when I received a visit from a European seismologist whose name I've forgotten. He'd come to study earthquakes in California. The two of us were poring over a graph when people came stampeding past us. The city was being rocked by a genuine earthquake—over a hundred were killed—but I hadn't noticed a thing."

He smiles at the recollection. Then he says, "What am I to do? Should I dance, get drunk, take drugs, sit on a tripod and utter prophecies over a crackling fire? Go insane?"

"You think even when you want to think of something else," she says.

"I can't help it—I can't forbid myself to. Thought is my

mistress, my dominatrix. There are times when I'd like to free myself from it. I dream of being someone else—I even dream of ceasing to think—but I still conceive of that someone else in terms of myself. I can't emerge from myself. Plato put that question in the mouth of sharp-witted Socrates, remember: 'What's the alternative?'"

"If you emerged from yourself, where would you go?"

"If I'm still inside myself, how can I tell?"

BEING READY IS ALL. Being there at the right time and being ready. Einstein was born in 1879, when electricity made its definitive appearance on the scene. It was the year when Edison perfected his incandescent lamp, which would soon light thousands upon thousands of homes.

Electricity, the mother of artificial lighting, was the long-awaited form of energy: insubstantial compared to coal, bright, practical, transportable and seemingly limit-less—a "fairy," as it would soon be called. It was destined not only to present physicists with new and unfamiliar questions, but also to facilitate a very large number of experiments unknown in ancient times.

When allied with magnetism it formed electromagnetic energy, and this brought with it other explanations of the world increasingly at odds with mechanistic ideas, which were obsolescent. A major battle—a decisive one—was joined in the nineteenth century: Maxwell versus Newton. Their theories were mutually rejective, yet both were beyond dispute.

So Albert Einstein arrived at the right juncture and his choice of physics was prescient. He doubtless sensed, very early on, that a great leap forward in knowledge was imminent, and that all would be decided in the very heart of matter.

Nor was he the only one. Dozens of scientists including Planck, Minkowski, Poincaré, Lorentz, Rutherford, Schwarzschild and Langevin were addressing the same problems, though they sometimes formulated them differently. Each in his own field and at the same or almost the same time could have discredited the classical concepts of mechanism on the one hand and electromagnetism on the other and achieved a breakthrough.

Einstein alone succeeded in doing so—why, no one knows. Perhaps because of an extra pinch of intuition, a view of things that was broader and certainly more synthetic from the outset, a less inhibited and more relaxed intellect, a love of mind games and adventures in thought. He was ready.

And this special intuition, this gift, would not wane in a hurry. In 1924, at the height of his fame, he received a letter from an Indian physicist named Bose. It concerned a point that had yet to be elucidated: the enumeration of the particles of light whose existence Einstein had confirmed and which would henceforth be called photons. Bose ventured to suggest that these photons should be studied not in accordance with traditional methods but as "indiscernible" objects obedient to the laws of quantum physics, which was then brand new.

Photons were not like other physical objects, Bose asserted. Their study called for another kind of physics.

Einstein leaped at this opportunity. Just as he had in the case of Brown and Planck, he developed Bose's arguments, boldly deducing that the properties of photons must also apply to certain constituents of matter—to other particles, now called bosons, which associate and collaborate unlike *fermions*, which derive their name from the Italian physicist Enrico Fermi.

By analyzing gases composed of bosons, Einstein demonstrated—and established for a long time to come— that quantum theory, one of whose founders he undoubtedly was, applies as much to light as to matter.

His mind was on the march towards a great dream: the unification of the universe.

- FOUR -

THEY HAVE BARELY returned to the study and closed the third door behind them—or so it seems to the girl—when the first door, the one leading to the waiting room, is flung open and Newton, in his buckled shoes and long black cloak, bursts into the room.

He's tired of waiting (it has been several centuries), he has things to say and reiterate, and he's looking grumpy and impatient. This special and general relativity that's messing up his gravitation, this quantum mechanics, these confounded strings and superstrings that have been ringing in his ears of late—what do they all amount to?

What is this light that's itself and not its opposite? And this energy that's claimed to exist everywhere, in matter of all kinds? Who is fooling whom? Will men of science be walking upside down from now on?

Three spatial dimensions—fair enough, he already knows about those, they're the basis of everything—plus a fourth, that of time, are held to be the constituents of something called "space-time"! What *is* this monstrosity? Was it really needed? Where was it found? How can space-time be envisioned or manipulated? It's a chimera, the offspring of a carp and a rabbit! Is Einstein going to explain it to him or not?

The two men speak English together, and it's clear that this isn't the first time they've met and clashed like this. Einstein presents his explanations politely but firmly, trying hard to spare the feelings of the illustrious Briton, who we sense is easily offended. He endeavors to get him to accept that human systems are provisional by definition. All of us must acknowledge this. Yes, even he himself—despite the reprieve he seems to have been granted like Newton and a few others, perhaps—will sooner or later have to resign himself to the supersession of his ideas. In part, at least. Special and general relativity, together with quantum mechanics and field theory, will one day join the great store of dead ideas—not that they're completely dead. A great idea never dies. A vestige of the imagination of an epoch or the vision of a superlative intellect—a genius, let's say—will always live on.

Newton seems somewhat placated by Einstein's use of the word "genius," because he consents to sit down for a moment. In a thoroughly amiable and conciliatory manner (he isn't sure he really thinks his own ideas are doomed to die some day), Einstein reminds him of the *annus mirabilis*, 1665–6, when Newton, on his own submission, underwent a sort of physicist's miracle and received a gift from the Almighty, an event similar to that which occurred in Einstein's own miraculous year, 1905, when he suddenly revealed himself to the world. There are years like that.

Einstein passes over in silence all the years Newton spent in the byways of alchemy, in which he became embroiled like so many others. He prefers to express his high regard for the theory of gravitation, with which he's well acquainted, and which, he says quite sincerely, still applies to certain aspects of the universe, if not all. A theory so widely accepted that it was, and still is, referred to as "classical." Newton would have deserved the Nobel Prize four or five times over, no doubt about it!

But how, in all seriousness, can one accept that the universe is composed of little balls that collide at God's sole pleasure? Newton miraculously sensed that objects could attract one another without necessarily coming into contact—without any push or pull being exerted by one or the other. He miraculously sensed (Einstein, who is laying on the charm with a vengeance, stresses the word "miraculously") the presence and effect of that decisive force: of the interaction known as gravitation.

But Newton assumed that the said effect exerted itself instantaneously from afar, whereas it has been proved innumerable times that the speed of light, even though it strikes us as very fast, is limited. So instantaneous action from afar is impossible. That much, at least, Newton must accept! After making such a fine start, he ought to go further! Why should we decline to advance boldly along the trails we ourselves have blazed? The theory of relativity,

Einstein declares, is merely a development and logical expansion of the great master's theory of gravitation!

Einstein also says (this is one of the passages the girl clearly understands), "After all, Isaac, I've never—never!—denied the real existence of the universe. I've simply stated that, for the time being, we can't find any absolute system of reference. That's far from being the same thing! Don't tell me I only like variables, when all I ever look for are firm invariants. How many more times must I tell you? The universe isn't absurd, it's relative!"

They embark on a technical discussion with the aid of graphs on the blackboard, giving detailed and, where necessary, recapitulated explanations of what they mean. Einstein displays extreme tact and patience. He's forever administering sugar-coated pills such as "You'll follow this very easily" or "As you yourself predicted" or "As your own work enabled us to discover."

The girl, who knows English and is listening intently, soon feels out of her depth. It's clear to her that she's witnessing an extraordinary encounter in which many fine minds would have liked to participate, a genuine summit meeting of universal physics. But what's the use? What can she gain from this privilege? How can she follow this dialogue? What, for instance, does "the perihelion of Mercury" mean to a layperson? When the two men refer to "energy," do they both mean the same thing? Einstein said he had trouble talking to Newton about elevators and airplanes, to cite his favorite examples, so what about "quanta," "neutrons," "electromagnetic force" and undulating "photons"—terms that sometimes cause the Englishman's hackles and eyebrows to rise?

Still preserving his smiling, benevolent manner and flattering Newton to the best of his ability, Einstein suddenly introduces him to his female visitor. He tells her that, in his own day, this great English scientist made some prodigious discoveries and established a very remarkable system. A system from which something, however, was missing.

"What?" Newton demands. "What was missing?"

"You noticed some flaws in your own reasoning," Einstein tells him, "but you didn't know how to account for them. You said, 'I feign no hypothesis.' But that wasn't true, admit it! Instantaneous action from afar was itself a hypothesis! You even invoked God, remember? You said that he had to intervene from time to time to put things right."

"Well, what of it?' Newton says.

"We can no longer accept such intervention."

"Why not?"

"Because—forgive me, my dear professor—the idea of God with a toolbox simply won't wash any more."

"Why not? Tell me why not! Surely you haven't renounced God?"

"Where the formation and organization of the universe is concerned, I'm afraid we have."

"What? You've all gone mad!"

"Perhaps."

"Wasn't it God who created the world? Created and organized it?"

"We no longer look at things in that light. These days we even prefer to eschew the word 'Creation.'"

"Really? Why?"

"'Because the world in which we live was not created in an instant. We still have no very clear idea of whence it arose or what we mean when we speak of a 'big bang.' But we're all agreed on one point: the world we know today took millions and billions of years to evolve, requiring clouds of stars, mixtures of sidereal gas, cascades of particles, and then, at a much later stage, a succession of very rudimentary life-forms that underwent gradual modification, many of them becoming extinct on the way . . . In short, it's a very, very long story. In your day, my dear and highly esteemed professor, you hadn't the wherewithal to conceive of such a time span."

"But if you've eliminated the divine creator of the universe, what have you put in his place?"

"Forces, as we call them."

"Forces like my gravitation?"

"Yes. We've retained it, among others."

"What others, for God's sake?"

As gently and patiently as ever, Einstein speaks of electrical and magnetic forces, whose functional principles Newton seems to assimilate quite quickly. He also shows the Englishman a flashlight and explains how to use it.

Newton takes the flashlight and turns it on and off. He's convinced—how could he not be? Besides, it probably isn't the first time Einstein or someone else has made him sample electricity. But he "tends to forget things," Einstein tells the girl under his breath. He plays the innocent and feigns ignorance as if electricity were a child's plaything, a minor force of no importance.

In any event, there it is, electrical force: Newton is holding it in his hand. He's acquainted with the magnetic kind, having manipulated magnets and observed the behavior of iron filings. Yes, that's another thing that might amuse a child. That's one point on which they concur: magnetism. Einstein tells him that, as a boy of only four, he was fascinated by a compass. Why did the magnetized needle persist in pointing in the same direction? What invisible force was it obeying?

Having listened to Einstein talking for a while, Newton quite readily admits that the two forces, when combined, make one: electromagnetic force. Thus far, all is well. At the end of the nineteenth century, Einstein tells (or reminds) him, most modern scientists believed that electromagnetic force would one day suffice to explain the coherence and workings of the universe.

"Abandoning gravitation?"

Einstein deems it better not to answer. How can he tell Newton that general relativity is merely a new interpretation of gravitation, considered henceforth as a curvature of space-time? How to pave his way to the new cosmos? How to introduce him to matter as newly conceived? Will he dare tell Newton that it is objects themselves—for instance the earth or any other heavenly body, or even a thrown stone or a falling apple—that modify space, or, to be more precise, space-time?

Could Newton understand those words?

Does Einstein dare to tell him that time itself is definitely just an illusion?

Innumerable worlds? Definitive uncertainties? Einstein prefers to avoid those matters, at least for the moment. But he can't avoid the atom, the thing of which people had spoken for so long and whose existence even Newton accepted, because it is there. It reappeared on the scene around the end of the nineteenth century, after a very long eclipse, and with it came the disclosure, manifestation and self-assertion of the invisible.

There they are—there in the realm of new objects and forces. Ever since spending time together, the two men have found mutual comprehension far harder when they come to tackle strong and weak nuclear energy. What are they, exactly? Newton fiercely knits his bushy brows whenever atomic nuclei and particles make their appearance. He yearns to shoo them away with the back of his hand, like troublesome insects.

Although they've spoken about those little things before, the Englishman is sorely perturbed by anything having to do with the atomic nucleus. One force to maintain atomic nuclei as they are, another to regulate certain disintegrations of particles—he finds the whole idea suspect. Why two forces? On what level do they operate? Where do they originate? Where do they obtain their impetus and the energy they require?

Besides, is all this really regarded as certain? Einstein says it is. He corroborates it scientifically, even though it isn't the aspect of physics to which he's most closely attached. "We're all in agreement now," he says, "those, at least, who have worked on it are. It's our form of certainty these days."

He goes on to say that their great dream would be someday to unite these four forces—strong nuclear, weak nuclear, electromagnetic, gravitational—so as to make one only. Thanks to the last-named, Newton would not be entirely unrepresented.

One force to govern the universe? Yes, says Einstein, everyone has dreamed of it. Newton did too, no doubt, in his day. All great scientists have dreamed of unification, of ending the disorder and mishmash, of finding the key to the universe, the philosopher's stone of physics. A force capable of being reduced to a formula—one of which all other forces would be facets. The infinite expressed in a few symbols. At present, for want of a better designation, this is termed the "theory of everything."

Newton shakes his head occasionally. He's looking incredulous and, above all, anxious.

"But why all this meandering to and fro?" he asks. "Why this compartmentalization? But for a few details requiring improvement, my system was magnificent! It worked! And God still played the leading role in it, to the greater joy of all! It was an excellent way of explaining the workings of the world by means of gravitation, and also of ensuring spiritual peace thanks to the exercise of true religion! What more did you want?"

"Things are different now, that's all," Einstein replies.

"We've got absolutely no quarrel with you, we're simply more demanding. Our instruments are more advanced. Your calculations, Professor, still work extremely well with heavy objects, for instance objects unknown to you such as trains and aircraft. Elsewhere, in other dimensions, the situation is different. Your theory of gravitation was only an approximation to mine. Some of your assertions strike us as incomplete or have quite simply been refuted by observation and experiment. You must admit that, especially as the world, or the universe if you prefer, is infinitely bigger and more complex than the one known to you and your friends. And that, forgive me, is none of my doing."

"What do you mean?"

Einstein heaves a sigh. Is Newton pretending not to remember, or has he genuinely forgotten?

It's easy now to tell that this isn't the two men's first encounter, and that principles formerly deemed eternal—for instance notions of space and time (it always comes back to them) and the perpetual stability of the universe—have recently foundered. Newton was convinced, absolutely convinced, that if some immeasurable force removed all the objects from space at a stroke, space with a capital S would remain, preexisting those objects and surviving them. It was the theory of the big empty box situated in time with a capital T, a time that would continue to "exist" even in an absolute void, even without manifesting itself—even if no object, whether gigantic or minute, were subject to it.

Einstein takes a different view of both space and time, but he has difficulty making himself understood, even making himself heard. Newton, who fails to grasp all the equations he's presented with, becomes exasperated. He takes off his wig, mops his bald, perspiring pate and quickly replaces it all awry. Then he asks, "Do you have any other explanations for all these wild hypotheses, these elusive and capricious particles, this vacuum that isn't a vacuum, this elasticated time of yours, this space that no longer possesses volume? I have a right to demand them of you, after all!"

"Yes, my dear Isaac, we can suggest some. Nothing has been definitely ascertained as yet, and you'll doubtless be able to assist us. Look, run your eye over these, I'm sure they'll interest you."

He picks up an armful of books and periodicals and hands them to Newton, nudging him gently towards the door.

"If there's anything you don't understand, especially in the mathematics, don't hesitate to come and see me again. I'm not a mathematician of the first order, but I'll do my best to help you. Here, take this too."

Newton allows himself to be shepherded out. Rather grumpily, he returns to the waiting room with a whole stack of books and papers, some of which he drops without even noticing.

ALONE WITH THE GIRL once more, Einstein addresses her as if he's continuing an uninterrupted conversation.

"It was, in fact, the impossibility of pinning time down—of measuring it exactly, let's say—and of delimiting space or even lending it a shape that led me, and not me only, to conceive of 'space-time' and think along those lines. Everything became, if not simpler, at least more acceptable."

"Can you tell me about it?"

"We don't have the time. Besides, I'm afraid we'd get into conceptual areas you'd find a bit beyond you."

"You don't think I'd be capable of understanding?"

"Forgive me again, but no. Not without a command of mathematical terminology, which you told me you don't have. In any case, you didn't come here to take a postgraduate course."

"How do you know?"

"I can tell by looking at you. You aren't going to become a physicist in one visit, and you know it."

"Why did Newton look so agitated?"

"Because he's afraid."

"Afraid of what? Of losing his illustrious aura?"

"Yes, partly. He was a very ambitious man and jealously guarded his resplendent status. He had completely eclipsed the work of his predecessors, the astronomer Robert Hooke for example. He never said a word about him, yet he owed him a great deal. He even had Hooke's name removed from the walls of the Royal Society when he became its president. Yes, he's every inch a man, very full of himself, very dictatorial and uncompromising. Suspicious, too. He accused Leibniz, without a shred of proof, of having purloined differential calculus from him. I've told him the same things seven or eight times and he feigns incomprehension, which is a symptom of pride. You were right about him. With men of that order, great men who have covered themselves in well-earned glory, there always comes a time when they mistake themselves for God—when the world must be constituted the way they say."

"Did that happen to you too?"

Einstein avoids the girl's eye for a moment.

"Did it?" she insists.

"Yes, it did. I admit that, and people have sometimes reproached me for it. I acknowledged that Minkowski conceived of space-time before me, well and good. But that Poincaré thought of relativity before me or, anyway, at the same time as me, and that Lorentz did so too—that I kept quiet about, it's true."

"Does it matter?"

"Not in the least. The universe couldn't give a damn."

"I thought as much. So Newton is scared of being knocked off his perch?"

"Yes, without a doubt. Who wouldn't be? But he's a scientist for all that. He knows the form. He too will bow to experimentation and evidence in the end. He'll have to admit that the speed of light is limited. No, I think there's something else involved."

"Like what?"

"Put yourself in his place. He has outlived his day by nearly three centuries. That's fantastic, in fact I think it's a record for a scientist. In my case, for example, it's been only fifty years, and I sometimes get the feeling I won't last much longer."

"So what are you being kept here for?"

"To work, of course, as I already told you. To continue our research and pass judgement on other people's, just as Newton does—with a total absence of good faith—on mine. We're engaging in spectral science. It seems that our equations, and mine in particular, harbor secrets so well hidden that they've eluded us. So we're all doggedly pursuing our research together, I and the people in the waiting room as well as the ones who write to me and phone me. The reprieve we've been granted doesn't amount to much in the great river of eternity. A tiny little victory over death—over physical death. The illusion of still being alive. And Newton, who has chosen to stand aloof from our discussions, must feel that his reprieve won't be extended again, that he's used up and out of the running—that he'll shortly dissolve and join the long, silent procession of useless shades, this time without any hope of returning. He'll be lucky if he leaves behind fifteen or twenty lines in the general history of the sciences."

"You too?"

"Of course. We obliterate each other in turn, and each of us retains a few crumbs of all his predecessors in his pocket."

"What about your famous equation?"

"Which one?"

"THE ONE EVERYONE knows: $E = mc^2$."

"Oh, that one. But there are others, you know, and they're just as interesting if not more so."

"Why do people only know that one?"

"Because . . . Yes, why do they, actually? Perhaps because they're lazy. Because it strikes them as the simplest, the most melodious, the easiest to remember. Because I formulated it in a postscript—absentmindedly, as it were. Who knows?"

He pauses for a moment, lost in thought.

"There are so many other things I'd like to talk to you about," he says at length.

"Like what?"

"The disappearance of the 'ether,' to which I contributed. We emptied the immensity of space, which was considered impossible. We rid it of that putative matter, invisible and impalpable but rigid, which had hitherto served to support light as water supports a wave, and which seemed absolutely indispensable. Yes, the ether evaporated at a stroke—it offered almost no resistance. I'd also like to talk to you about curved space-time, which we've barely touched on, and about the granular composition of light, which is our supreme messenger—which we continuously emit or reflect and thanks to which we see each other."

"Do I emit light?"

"Yes, you receive it and send it back to me, or I wouldn't be able to see you. We see objects only because they emit light like the stars, or because they're illuminated like the planets and send back the photons they receive."

"Does the light I reflect take a certain length of time to reach you?"

"Of course. Like all light."

"It's very quick."

"But not instantaneous. I don't see you the instant you speak to me or stroke your hair. I see you a little bit later, by which time you've already changed. You aren't the same anymore. I see you as younger than you are by some minifractions of a minisecond, and the same applies to all vision. Remember what I told you, out there among the stars, about simultaneous occurrences. As you appreciate, the word 'simultaneous' possesses no more meaning than the word 'measure.' Whether we're talking about an interval of a light-year or a nanosecond, everything is discrete. You are at a distance from me, like a star."

"Is there light everywhere?"

"Even in dense shadow, but extremely diffused and subdued. It's the prime mystery, and you probably know how fond I was of the mysterious—how fond of it I still am."

"You wrote about it."

"Yes, in what people call my profession of faith."

She quotes from memory, "'The most beautiful experience we can have is the mysterious.'"

"Yes, that's it, more or less. They even made a recording of the text as if I were a café singer. But it's true that all sublime activity, in science as in the arts, is based on the mysterious. The mysterious is the prime mover of all mental activity."

"But the arts obscure the mysterious," she says, "whereas the sciences seem to make it their mission to dispel it."

Einstein disagrees. It's true, he says, that the arts tend to seek obscurity—that's what they're there for—but they can also arrive at resounding truths that are personally experienced and felt by a very large number of people, whereas scientists are forever roaming from one mystery to another. And, when they think they've discovered some kind of new light, they don't know how to make it acknowledge that discovery. Darkness resists, digs its heels in, builds itself new lairs. Hence the liking scientists cherish for the inexplicable, and hence their attraction to the unknown. (A perverted taste? Who knows?) It's as if vast territories of a potentially hostile, lethal nature have only been awaiting their arrival to reveal themselves ever since the world began.

And to think that there are still some who feel a nostalgia for magic, for the great, encoded secret preserved by initiates, for signs that conceal things, for symbols and numbers! They have forgotten that in *The Tempest*, at Shakespeare's behest, Prospero consigns his magic book to the ocean bed forevermore. His charms and spells are "o'erthrown," tossed overboard. He abandons them and goes home. A huge page turned at the beginning of the seventeenth century. Copernicus was dead and Galileo hard at work, Descartes had already been born. A brave new world was in the offing. Elves and goblins were disappearing who knows where, fairies were in hiding. The last witches were being burned at the stake by frenzied fanatics.

However, even supposing we occasionally miss the magic spells of the ancient world, how can we fail at least to glimpse the boundless territories now on offer? How can we fail to see that the world's great enchantment is only just beginning?

"I could also talk to you," Einstein goes on, "about the speed of light, which I saw fit to make a sort of *ne plus ultra*, a fixed and unsurpassable measurement."

"Why unsurpassable?"

"Because if objects moved at a speed exceeding that of light, and if one multiplied the square of that speed by the mass of those objects, as suggested by the equation you chose to quote, the objects would be of infinite mass. Which, as you'll agree, is inconceivable."

She briefly speculates on what an "infinite mass" would be like, turning the two words over in her mind. And it's true: she can neither visualize nor conceive of such a thing. Her thought processes stop short of the unthinkable, which is natural enough.

They do, however, suggest that there may be other dimensions, other universes. She isn't too sure how to put this.

"Do other universes exist?" she asks bluntly.

"Some of us claim so. Speaking for myself, the one we know is quite enough for me."

"But all the same! Tell me! Several universes, just imagine!"

"Well, yes, perhaps. We do have certain pointers in that direction. No proof, but it's quite possible that the universe we see isn't the only one. Or rather, it's the only one but existing in multiple forms and provided with supernumerary dimensions. Only one of those forms appears to us and we think it's unique. We believe it's unique because it's the only one we can see, touch, observe. Yet another illusion that deceives our eyes and senses, even our minds."

"*You* say that?"

"Let's be clear about this. If universes entirely discrete from us do exist, they're inaccessible to us, so they might as well not exist. If they're linked to our own, on the other hand, they're another form of it, so there's only one universe."

"But are there types of matter other than ours?"

"That's beyond dispute. Several of them, actually. I didn't say that, others who came after me did. First they discovered antimatter, which was relatively easy, then dark matter, and then yet another 'substance'—we don't really know what word to use. A repellent substance, not attractive like ours, it occupies over 70 percent of the world's total mass! It's preponderant but unknown! In this context people go so far as to talk of 'vacuum energy' and 'dark energy'—even, like medieval alchemists, of 'quintessence.' I merely prepared the ground with my space-time palaver. The last thing I expected was such a bombardment—or such a metamorphosis, if you prefer. I sometimes feel that this unnameable matter is not so very far removed from my 'cosmological constant.' At other times I continue to question myself. But the fact that we distinguish several types of matter doesn't mean that there are several universes.

"Nowadays," he goes on, "when we look at this complex universe through the powerful artificial eyes with which we've equipped ourselves, we see that it's a kind of multidirectional effervescence in which any event can erupt, fall back, rebound or disappear in one direction or another. From now on, neither time nor space—nor even space-time, perhaps—is in command of the ship. Computation and reasoning are at an end. We may possibly have transgressed the bounds of what is intelligible. We're tottering on the frontier."

"And your equation?"

"Ah yes, my equation. Well, it was an idea that had been prowling around all over the place, like all ideas. I finally formulated that idea after two years' hard work."

"What idea?"

"That matter and energy are the same thing."

"Meaning what?"

"It was a question of relating two concepts that had hitherto seemed entirely discrete: matter and energy. It used to be said that energy was required to set matter in motion, and everyone looked for the source or sources of that energy, for the forces inherent in this or that, in wind, in cold, in heat, in God. Look: there's energy in all matter. In this, for instance"—he picks up a sheet of paper and lets it fall to the floor—"or in a piece of wood or iron. It's everywhere and in everything."

"On what level?"

"On an invisible level, of course. On the level of an atomic nucleus. That's why no one could find it."

"In what proportions?"

"Unimaginable proportions devoid of any measurement in common with the dimensions of an atom. An abundance of energy in the smallest volume imaginable. In order to account for a certain number of phenomena, therefore, I postulated that energy, which we call E, is equal to matter—or mass, if you prefer—multiplied by an enormous figure: the square of the speed of light, which I call c."

"Could we make use of that energy?"

He avoids her eye and pauses for a moment or two. Then he says, "We already have."

"You mean it's nuclear energy?"

"Of course. Energy obtained by means of nuclear fission. By splitting the atomic nuclei of certain metals."

"And setting off a chain reaction?"

"Exactly."

- FIVE -

ANOTHER BRIEF SILENCE. Einstein keeps his eyes lowered, fingers drumming soundlessly on the desktop.

The girl notices his agitation. Or irritation.

"Does it upset you to talk about it?" she asks.

"Yes and no. Since you made some allusion to the fact when you came in, you must be aware that people have accused me of being the originator of nuclear weapons. Of being responsible for Hiroshima."

"And it isn't true?"

"If I was responsible for that horror, it was only very indirectly, believe me. It never occurred to me, even for an instant, that those consequences would result. I didn't believe in atomic fission to the very last—I drove the idea from my mind. I wasn't even kept informed of how work was progressing, I assure you. I'd campaigned for peace all my life, for an organized and well-secured peace. And besides . . ."

"What?"

"I'm a Jew, did you know that?"

"It's hard not to."

"I'm not a practicing Jew, far from it, and I'd never attached any great importance to my origins although I come of Jewish stock. I was about forty years old when I first felt Jewish. That was when the harassment and vilification began. I regarded myself as a German who had become a naturalized Swiss to avoid military service until . . . You hear that?"

We do, in fact, hear the approach of a German military band, the tramp of boots, words of command, the roar of engines, cries of hatred, Hitler's strident voice.

"Life can take a sudden, stupefying turn—one that it's hard to forget. It was a terrible shock. Come and see."

He leads the girl over to one of the five doors and cautiously opens it a few inches.

They both peer through the crack.

Visible beyond the door, as though projected on an outsize screen, is some archive footage accompanied by violent sound effects: Nazi parades, attacks on Jewish shops and synagogues, volleys of stones, shattered windows, books being publicly burned.

They step forwards. Einstein positions himself in front of the girl, and she sees his long white hair flutter in the wind from the street. The images cease to be in black and white. Colors appear, together with an impression of proximity and reality; they can feel the heat of the bonfire and smell the smoke. The printed paper burns with difficulty and has to be doused with gasoline. Voices seem to ring in the girl's ears—she even feels she's being shoved and jostled, feels splinters of glass showering her face and shields it with her hands.

"You see?" Einstein tells her. "My own books are in that batch, along with those of Freud, Thomas Mann, Proust, Stefan Zweig and many others. They were burned. No, you aren't dreaming: they were tossed into the flames and burned. A barbarous, preposterous act. Totally idiotic. A fundamentally superstitious act too, as if Freud and I were sorcerers, creatures from hell whose misdeeds only fire could obliterate. As if ideas could be destroyed by burning some paper."

Turning to her, he can tell from the look on her face that she's frightened. She's alive and young and afraid for her physical safety, her life.

"Let's go back," he suggests.

She withdraws, but only slowly, as if something about the past that envelops her—about this firsthand contact with undying hatred—is holding her back despite the danger.

He closes the door, but we can still hear the crackle of flames, the sirens, the cries of pain and fury, the gunshots, the smashing of glass, the rumble of armored half-tracks.

Back in the office they pause to regain their breath. She asks if he was physically assaulted.

"It was mainly verbal abuse. They claimed that my theories were absurd and dangerous. Imagine, they spoke of 'Jewish science'! They put death threats in my mailbox. At home in Germany they said I was a liar and a traitor. They said I hated the Germans!"

"Were you very well-known by then?"

"Yes, and had been since 1919, when Eddington produced his verifications. That amazed me more than anything else. I was constantly pursued by journalists and photographed almost daily, wherever I went. People lay in wait for me—they stalked me like some strange beast. But why? I asked myself that question again and again without ever managing to answer it. How many people had read my four articles in the Annalen, a very specialized periodical, and how many had understood them?"

"As an Englishman, Eddington had been an enemy of Germany during the war, hadn't he?"

"Are you implying that we represented international reconciliation? That the two of us were a symbol? You're mistaken, there's another side to that coin: Eddington was an Englishman, ergo an enemy, yes. But many 'patriots' among my fellow countrymen—forgive me, it always makes me feel slightly queasy to use that word—stated that because the war had ended in defeat no enemy could be trusted. It was out of the question. So I was not only well-known but suspect."

"And you were watched?"

"Yes, all the time, take my word for it. The craze for surveillance originated during those years. I was the last person to have suspected such a thing because I credited myself with no political importance, as you can well imagine, but later I had proof of it. Detailed reports were compiled about my trips abroad, for instance to the newly established communist countries, first by the German secret services, then by the Americans. It was claimed that I was on their side—that they used my study as a safe house! Not that I realized it, but I was living between the dreary pages of a spy novel. I was dogged by that situation throughout my life, did you know?"

"How come?"

"Later on, in the United States during the 1930s, they wanted to know if I might by any chance have retained links with Nazi Germany and Hitler's henchmen, the ones who had burned my books—and that at a time when the Nazis considered me a bandit, a dangerous madman! Some American women who styled themselves 'patriots' opposed my admission to their country on the grounds that I was German, Jewish and probably a communist! They called me 'worse than Stalin'! Honestly, I was spared nothing. After 1945 the Americans seriously wondered if I was conspiring with the Russians! With the Reds! A secret FBI report had advised against employing me on confidential projects. It actually stated, in the very year I was granted American citizenship, that I couldn't have become a loyal American in so short a time!"

He bursts out laughing again—laughing at his own checkered existence—so loudly that it makes his mustache twitch. Then his laughter subsides and his mustache comes to rest.

"The longtime director of the FBI," he goes on, "that detestable Mr. Hoover, stopped at nothing, forgery included, in his attempts to convince public opinion that I ought to be expelled from America's sacred soil. I was branded the head of an anti-American network, a 'Red front' dreamed up *in toto* by the intelligence services. Even my faithful Helen Dukas was placed under surveillance! I could go on about it all night, but it would tire me out and disgust me, truly it would. I loathe dwelling on the contemptible aspects of the human race, and in that regard we haven't progressed a millimeter."

"Some people even say the opposite."

"The opposite? No, no, I don't think so. We couldn't become worse than we are, I can't believe that. That advances in our knowledge of the universe lead to moral regression? No, I don't believe it. Certainly not."

"Nobody claims that science necessarily makes things worse," the girl goes on, suddenly feeling on weak ground. "It's simply that the two tendencies are almost concomitant: the more knowledge, the more horror."

"But the horror is the same as before. It acquires greater resources, that's all."

"Thanks to you."

Einstein stares at her in silence for a moment. He seems to have lost all desire to smile.

"Thanks to me?" he says. "You truly believe that?"

She gives a little shrug and doesn't reply.

He looks away.

GERMANY AT THE beginning of the 1930s. The Weimar Republic, which came into being after the last of the Hohenzollerns, Kaiser Wilhelm II, abdicated in 1918 and has never stopped moving to the right since that date, is already in its death throes. The Wall Street Crash of 1929 has devastated the world economy and affected Germany in particular. Galloping inflation, infectious loss of confidence, pathological resentment, multifarious accusations and a fierce desire for revenge, mainly in respect of the "shameful" Versailles Treaty to which Germany was compelled to submit at the end of the war—all is in readiness for the door to open and the monster to make its entrance.

In this increasingly oppressive atmosphere of doubt, tension, suspicion, denunciation and latent ambition, Einstein (who tells his friends that Hitler derives his strength from Germany's "empty belly") becomes famous overnight and, even though he fails to understand why this is so, strives to exploit his fame on behalf of causes he considers just.

He makes appearances all over the place. In 1930, for example, he delivers the inaugural address at Berlin's Radiophonic Exhibition. He is received at the Chancellery (sometimes with his friend Planck); he rubs shoulders with government ministers; he attends and quickly becomes the star of the Solvay Congresses, which, at the instigation of a wealthy Belgian industrialist, have since 1911 brought together the greatest names in science; he delivers more and more lectures; he plays the violin in synagogues, his own form of protest at the insidious growth of anti-Semitism, which is already poisoning Germany and Austria; he participates in fundraising; he signs petitions; he campaigns to the best of his ability for a new education system; he believes in the social utility of science; and he often attends meetings of the International League for Human Rights.

In 1922, during a visit to Japan (he receives a telegram on the voyage informing him that the Swedish Academy has at last awarded him the Nobel Prize for physics), silent crowds follow him through the streets like a kind of prophet. Public opinion proclaims that he has succeeded in plumbing certain secrets of the natural world, and that, even if those secrets appear to be incommunicable, he is a hero belonging to the whole of humanity—one who personifies its latest victory over the dangerous mystery surrounding it. Those who speak of him almost obligatorily employ the word "genius." Einstein is now an official genius.

An observatory built at Potsdam bears his name. Some years later his likeness will be carved on the portal of the interdenominational Riverside Church, New York, in company with saints and important figures from every era including Moses, Kant, Milton, Darwin, Descartes, Beethoven and Thomas Aquinas. When welcomed there by the minister, he's astonished to find himself the only living person represented among the sculptures on the doorway. His reaction: "I shall have to be careful what I do and say for the rest of my life."

Einstein is privately amused to have wound up as a Protestant saint when he's never succeeded in becoming a Jewish one.

"Did you know Freud?" the girl asks.

"Yes, during those years. He was older than me and

extremely well-known too. I used to call him *Der Alte*, the old man. He said I was as hopeless at psychology as he himself was at physics. Perhaps that's why we got along so well."

"He was Jewish too?"

"And a non-believer just like me. Remote from any form of religion and rather skeptical, I think. He likened religion to a collective neurosis or infectious disease—something like that. He also distrusted nationalism and bigotry of all kinds. In his last book he even tried to deprive the Jews of their founding father—to convince them that Moses was an Egyptian whom the early Jews had assassinated! They didn't believe it, of course. I sympathize with them."

"You thought the two of you could achieve something?"

"At least we tried. With no illusions on his part or mine."

EINSTEIN HAD BEEN violently, instinctively horrified by war ever since his childhood. He was a "pacifist," to use the contemporary description. He detested weapons, military parades and fanfares, he didn't believe that the world should be ruled by brute force and discipline, he had no faith in the virtues of victories, surrenders or treaties. He was opposed to nationalistic sentiments of all kinds, he had even called for an inquiry into German war crimes during the First World War, he cherished a desire for an international world authority and would continue to do so throughout his life, believing (though he was sometimes aware of his own naïveté) that this alone could put an end to war. It distressed him to see a large proportion of humanity exchange religious mania for that of nationalism.

During the First World War he refused to sign an "Appeal to the Civilized World" *(Aufruf an die Kulturwelt)*, whose ninety-three prestigious signatories included his friend and colleague Max Planck and Max Reinhardt, the distinguished Berlin theater director. This wartime appeal urged support for Germany and unreservedly sang the praises of German militarism, declaring that "the German army and the German people are but one."

He even signed a "Counter-Appeal" calling for an immediate cessation of the war and international understanding. According to the writer Romain Rolland, who met him on neutral Swiss territory, he went so far as to hope for an Allied victory and the final defeat of Prussia.

He was absolutely delighted by that defeat and the fall of the Kaiser, confidently hailed the advent of the Weimar Republic and campaigned on behalf of the new Germany. In Berlin, during a rather lively demonstration, he managed to secure the release of the university authorities, who had been detained by some students.

He took the floor at every meeting he attended—and he went to more and more of them—but his feeble voice was almost inaudible in the pandemonium despite his reputation. So he wrote Freud a letter posing the question that haunted him, one to which physics could provide no answer: Why war?

How could the masses allow themselves to be whipped up into a frenzy of self-immolation? Was there some means of controlling people's psychological development so as to render them more resistant to the psychoses of hatred and destruction?

Einstein addressed those questions to the man he called a "great expert on human instincts."

Freud sent him a long letter in reply, and their two texts were published in 1933, just before Hitler was elected. Freud affirmed his belief in the existence within us of an urge for hatred and destruction. Urges of this type were the contrary of erotic urges, which aspired to conserve and unite, but both these contradictory urges were equally essential. Neither of the two could ever express itself in isolation and each was always alloyed with a little of the other, so they could not be simplistically reduced to good and evil.

According to Freud, all living creatures had within them an urge to reduce life to the state of inanimate matter. But it would be useless to claim to be able to suppress these destructive human proclivities completely. War, Freud told Einstein, had been part of us for a very long time. It had accompanied our advance towards civilization, which had inevitably entailed psychic transformations. We had progressively lost our instincts and curbed and controlled our impulses. That was why we had come to regard war as intolerable.

How were we to rid ourselves of it? Freud admitted he

couldn't say, but he did assert that "all that conduces to the development of civilization also militates against war."

The two most celebrated Europeans of their day had done their best. They had added a drop of clear water to the muddy river that was swelling and expanding in all directions.

Too late, however. Their voices remained unheard. The human mind at its most acute—history has witnessed this often enough—was overwhelmed once more by poisons of its own devising.

WHAT ARE WE TO SAY when a mind roams the universe unceasingly, when it sallies forth and strays into immeasurable wastes, into an infinity of possibilities, and turns for a moment to focus on the earth's minuscule disputes, on reciprocal demands for territory and possessions, on mutual insults and threats hurled from either side of an imaginary line, a mountain range, a river, a customs station—on the narrow- and mean-minded nationalism that paralyzes us?

How is such a mind to react on returning from the stars?

It undoubtedly has difficulty in involving itself in our petty terrestrial conflicts. It views those little matters from too far away. The stars are no one's property.

Nonetheless, that mind belongs to an individual who also, whether he likes it or not, belongs to this planet. He was born here, not out there, he hails from a particular place, he has been invested from childhood with a language and education, sentiments and ideas. He has grown accustomed to a certain way of speaking and dressing, behaving and eating. He has grown up and developed in a particular corner of this earth. When the race to which he belongs is threatened and despised—even crushed, despoiled and expelled—how can he fail to react, temporarily forgetting about infinity and clinging with all his might to the endangered grain of sand that is his?

Unknown at first, then famous, Albert Einstein did his utmost on behalf of the thing we call peace, but to no avail. Two successive spells of worldwide carnage occurred during his lifetime, and, by reason of his research, he found himself in an extraordinary, historically unique position. Even as his pleas for peace were fading and dying, he devised a musically pleasing equation that would very soon lead to destruction.

"Is it true," the girl asks him, "that you once said you'd survived two wars, two wives and Hitler?"

"I don't remember, but I may well have done so."

FROM 1930 ONWARDS he was a visiting professor at Princeton University in the United States. But he had already been to the New World before. He made his first trip there in 1921, accompanied by some friends and his second wife Elsa, a cousin whom he had married in 1919, the year of his divorce and rise to fame.

Together with the militant Zionist Chaim Weizmann, he helped to raise funds for a Hebrew university in Jerusalem (which opened in 1925). Initially hesitant because of his abiding mistrust of nationalism of all kinds, he was prompted by the growth of persecution in Europe to espouse the idea of creating a Jewish state. He also accepted, albeit reluctantly, the idea of traveling to Dollarland. (Much later, in 1952, Ben-Gurion offered him the presidency of the new state of Israel, which he declined on grounds of incompetence.)

Even by the time of his first visit to America, legend had already insinuated itself into his daily existence. It is reported that the liner's crew took turns standing guard outside his cabin to prevent anyone from disturbing him at his work during the voyage.

"Were you well-known in the United States too?"

"Even before I got there. It's a country where people adore celebrities, you know. I was wearing a label that said 'world-famous' and I didn't even know why. Or rather, the people who greeted me with bouquets and medals didn't. I could have made a fortune out of advertising. Look at all that!"

He points to a stack of photographs, books about him, newspapers, brochures, requests to use his picture in advertisements for cigars or pens. Thoroughly amused, the girl roots around in these documents and knickknacks (which include an Einstein saltshaker and an Einstein letter-opener).

"It still goes on," she says. "I thought about putting on an Einstein T-shirt before coming here, but in the end I didn't dare. I told myself you'd take it the wrong way."

"You must be joking, I've even worn one myself."

"It's extraordinary all the same," she says. "World-famous, and nobody knew why."

"When we came ashore in New York I said to my wife, 'No one on earth deserves such a reception. I feel as if we're frauds and we'll wind up in prison.'"

"So none of the people who fêted you could understand your work. Is that true?"

"Hardly any of them. The journalists couldn't even say what I was working on, and not only in America. It was like that from the start. When Eddington announced the results of his work in London in November 1919, do you know who reported it for the *New York Times?*"

"No idea."

"The golf correspondent!"

He roars with laughter (the girl is getting used to these outbursts).

"I was taken to be everything I detest," he goes on. "A species of mysterious visionary, an abstruse guru, a sage with the secrets of the universe tucked in his pocket. But what the Americans found most intriguing of all was the fact that I often wore no socks."

"Why not?"

"It wasn't deliberate in any way. I didn't wear any, or wore only one, quite simply because I'd forgotten. I get dressed in the morning, I put on my first sock, an idea occurs to me, I get up to make a note of it—and good-bye to the second sock. Nothing extraordinary about that—pure absentmindedness. Besides, I don't know if you've noticed, but your big toe always ends up making a hole. There's no point in darning them."

"People don't darn socks anymore."

"They did then. I gather they're thrown away these days. Anyway, I came to realize that, as long as it's not too cold, a person can exist perfectly well without socks. You ought to try it. There's nothing more absurd than a sock, with the possible exception of a tie. Besides, can I tell you something?"

"Please do."

"I grew to like it. Here, look."

Using both hands, he hoists his trouser legs to reveal bare ankles and an expanse of white skin above his sandals. He adds that the physicists of those days, like all scientists, still plied their trade in shirt and tie. They wore cufflinks when handling their test tubes. It was as if the austerity of middle-class attire, invariably dark in color, was essential to rigorous cerebration.

He, for his part, having at first been obliged to dress like everyone else, forgot by degrees to do up his cuffs and thread his belt through the loops. The wider his mind ranged, the more his body assumed a vague, disheveled appearance.

"It's too late to change now," he says. "And anyway, wing collars are a rarity these days."

He describes having met the astronomer Edwin Hubble in the United States. Hubble, who headed the

Mount Wilson Observatory, persuaded him to accept that the universe, which he had initially assumed to be static, was expanding—a correction he readily admits. The girl tells Einstein that Hubble has given his name to a celebrated telescope that photographs the stars.

"I know," he says, "I've seen the pictures. They're very beautiful, but they're only pictures. Talking about celebrities, do you know what Chaplin told me?"

"No."

"He said he was very well-known because everyone understood him, whereas I was even better-known because *no one* understood *me*. There's one minor difference, I told him. When you go out and mingle with people you take off your mustache and few of them recognize you. Without a mustache your face becomes a protective mask, whereas I'm obliged to keep mine on."

"Has your fame been useful to you?"

"To my work, absolutely not. On the contrary, it hampered me and wasted a lot of my time. My dearest wish was to be able to think in solitude, in the background."

"And apart from your work?"

"It enabled me to live quite comfortably, no more. I was offered fictitious academic posts at Princeton and elsewhere. I gave some lectures, wrote some articles. But money didn't interest me too much. I always lived rather modestly. I—"

He breaks off and falls silent for a moment or two, as if he's listening to something.

The girl listens too. She turns to look at one of the doors. When she looks back at Einstein he's gone.

She peers in all directions. No more Einstein. He has vanished from one moment to the next. She's alone in the big study with the mutating decor. For an instant, as if overcome with fear, she seems tempted to leave. After all, this indeterminate place may conceal countless lurking dangers, unexpected flaws and pitfalls. Remembering the sci-fi stories in her grandfather's barn (travelers sucked into spatio-temporal crevices), she grabs her tape recorder, her shoulder bag and the few notes she's taken and makes for the door to the waiting room.

Then she stops short, possibly reflecting that it's a shame to leave so soon, and that she won't get another such opportunity. She reaches into her bag and takes out her mobile phone, but it's no use. No signal, not even a dial tone. She presses several buttons, but it isn't working.

Not particularly surprised, she sets off for the door again. Should she try to have a word with Helen Dukas before leaving? Perhaps. She takes several steps towards the narrowest of the five doors.

Then she hears, coming from outside the study, a hubbub of voices, doors slamming, glass splintering. Curiosity prompts her to change direction. Abruptly, she strides over to another door and opens it.

She finds herself in a conference hall somewhere in Germany in 1931 or 1932. Albert Einstein, who's in the chair, is trying to speak, but catcalls and whistles ring out on all sides, feet drum on the floor, an inkwell goes flying, chairs get smashed. Furious voices call for the extinction of Jewish science—a contradiction in terms, they bellow, because Jews are incapable of discovering the truth or even of acknowledging it. Their minds are contaminated, perverted, closed to the truth, as they have often demonstrated since the beginning of their lamentable history! Philipp Lenard, winner of the Nobel Prize for physics in 1905, has even written, "Jews are manifestly incapable of comprehending the truth, unlike Aryan scientists."

So down with Jewish equations! Down with relativity, curved space-time and other anti-German aberrations! Let's make a clean sweep of them all and save our Aryan Fatherland! Quick, to the stake with these self-styled Jewish scientists, these mendacious individuals, these prophets of universal corruption, these destroyers of wholesome sentiments!

Einstein protects himself as best he can, but he's compelled to gather up his notes in a hurry and leave the platform, half-shielded by one of the organizers.

The girl is there in the hall. She can smell the cigarette smoke, see the spectacle of pure hatred that presents itself to her gaze once more. She watches and listens to this person and that, hears the angry shouts that aspire to be reasoned arguments, almost understands the German in which they're uttered.

Unwilling to linger, she retreats a few steps, backing away towards the door by which she entered, and finds herself in the study once more. She quickly shuts the door. The hubbub in the meeting hall still carries to her ears, but more faintly. It's beginning to recede into the past.

Then she hears Albert Einstein's voice behind her.

"Now do you understand why I left?"

- SIX -

FOR LEAVE HE DID. Or rather, he never returned.

In 1933, when the Nazis officially assume power in Germany, he learns that his Berlin apartment has been ransacked several times, like his lakeside chalet at Caputh. Jews have been arrested, political opponents imprisoned or executed. Planck advises Einstein to resign from the Prussian Academy of Sciences, which he does. Should scientists become involved in politics? Are they even entitled to express their views? This is vehemently debated. "Science doesn't think," Heidegger will say a little later, whereas Einstein, like some others, is always referring to "thought." Are they speaking of the same mental activity? Can scientists prevent themselves from thinking? Must they cut themselves off from all observation of and reflection on the world they're supposed to describe and explain?

Scientists of Jewish origin are beginning to leave Germany. Ideas are clearing out.

Einstein is accused of high treason. It is said (though

never proved) that there's a price of $5,000 on his head (an inordinate sum, he's supposed to have remarked). He goes to England, chairs meetings when invited to do so, dines with Churchill, renounces his German nationality, declares that he will never return to his native land while the Nazis are in power, and initially settles down at Coq-sur-Mer in Belgium. A German newspaper denouncing his crimes prints a front-page photograph of him captioned "Not yet hanged!" Policemen assigned to protect him patrol his garden and sleep on his stairs.

At the invitation of Abraham Flexner, who heads the recently established Institute for Advanced Study at Princeton, he leaves for the United States. On arrival he cunningly eludes the crowds and the photographers. He tries to pass unnoticed. Living and working in the little university town that will bask in his reflected glory, he continues to occupy a modest house there until his death in 1955.

The equipment he requests for his very simple study includes a large wastebasket—to hold all his blunders, he says. A voluntary, exemplary exile, Einstein never returns to Europe.

"Did you miss Germany occasionally?"

He shrugs, makes no real reply.

"Your nice house at Caputh," she persists. "The lake, your friends . . ."

"I don't care to talk about it."

She relapses into silence. Despite her youth, she undoubtedly senses a trace of lingering nostalgia, an ill-healed, still-smarting wound. He may have changed his perception of the world, but he hasn't changed his country. He may have transformed time, but he hasn't managed to defy history. Germany remains his cruel parent.

Even if, in the strictly scientific domain, the great age of inspiration seems over; even if he's challenged at the Solvay Congresses, in particular by the bright new brains of the Copenhagen school headed by Niels Bohr, who can be obstinately dogmatic; even if he himself sometimes recognizes, towards the end of his life, that he has been mistaken in his search for a unified theory and his dogged insistence on an order and reality independent of the observer, a causality such that no physical influence can propagate itself at a speed exceeding that of light—even now his reputation remains intact. His face and silhouette are known and recognized throughout the world. People maintain that he's a genius without being too sure, even after all these years, why he deserves that appellation. Awards and tributes descend on him like rain. He cannot go anywhere without receiving honorary doctorates and gold medals, statuettes and diplomas, even the occasional check.

He is photographed on each of his public outings, so much so that he once declared his true profession was that of a photographer's model. In 1948, on emerging from the hospital after a brief admission for surgery, he sticks out his tongue at one of the photographers who are pestering him, and this impish image remains a twentieth-century icon, even today.

"On one occasion," he says, "yes, my fame did come in handy, though not to me personally. I'm not even sure it proved of use to anything or anyone, but it certainly made use of *me*."

"Are you talking about your letter to Roosevelt?"

"You know about that?"

"I did a bit of homework before I came."

He falls silent for a moment. It isn't a palatable memory, far from it.

"Did the fact that you were Jewish play any part in your decision to sign that letter?"

"Yes, perhaps. Secretly. By 1938–9 we already knew what was going on in Germany. We were aware of the horrors in store, it was impossible not to be. One had only to read their proclamations, their speeches predicting imminent extermination and calling for murder as well as total war. Hysteria was taking possession of the speakers' platforms and making their microphones vibrate. Their strident voices might have been issuing from a single throat. The big arms dealers were prodding them hard in the back. As for the rest of the world, it kept its eyes and ears shut."

No one, or almost no one, could understand the new ideas Einstein set forth.

No one could understand—or was prepared to admit—that the German nation had entrusted its fate to a criminal lunatic.

"Did you know that the Nazis were working on an atomic bomb?"

"We suspected it without being sure. They were technically capable of it, because they had some excellent physicists like Hahn, Strassmann and Heisenberg, among others. Were they providing themselves with the necessary resources? We didn't know at that stage."

"Was Niels Bohr convinced of it?"

"Yes, certainly. Heisenberg came to see him in the middle of the war. Niels had already been warned by one of his associates, a woman of Jewish origin whom Hahn had informed. Heisenberg paid a secret personal visit to Denmark, an enemy country, at the risk of being charged with high treason. But what did he tell Bohr? Did he disclose that Germany was devoting itself to perfecting a nuclear weapon? Or did he merely, as some believe, talk about nuclear reactors for peaceful use, and Bohr extrapolated? Frankly, we just don't know. Our information remained confused, sometimes even contradictory. I myself didn't believe it, I assure you. I persisted in my naïveté. I'd never imagined that anyone could use my theories in an attempt to destroy the world. A Czech student had tried to show me that it was feasible in Prague one day, but I didn't even listen. When people asked me the question I always said it was impossible—idiotic."

"So you shut your eyes too."

"Perhaps. I didn't believe that one of my equations could unleash the Apocalypse. They hadn't been conceived with that in mind, of course. They were research, pure and simple. It's impossible to reject compelling evidence. All you see is its good side, the increment in knowledge, in the truth of matter itself. That's what you're there for. Even factually, even technically, I just couldn't accept it. In order to transport an 'atomic bomb' designed to flatten a big seaport, for example, we calculated that an enormous, remote-controlled ship would be required, but how was it to be done? Would a vessel of that kind—I mentioned this in my letter to Roosevelt—be able to cross the Atlantic and reach New York without being detected? To me such an attempt at annihilation seemed unimaginable—beyond the capacity of any human mind."

"Yet people had been dreaming of it for so long . . ."

"You're right. Our dreams are quite as often destructive as constructive. Sodom and Gomorrah, the fire from the sky, the end of this corrupted world, the fall of the Tower of Babel, 'Babylon is fallen, is fallen, that great city . . .' And further afield—I was told this in India—the most ancient writings speak of a total weapon known as Pasupata, which can annihilate all life on earth."

"All life?"

"According to an old poem, even the grass on the battlefield trembled with fear."

"Why this terrible desire to destroy everything?"

"Ask Freud."

"Could I meet him too? Has he also been granted a reprieve? Is he around somewhere? Is there a chance of my seeing him?"

"I have no idea. I haven't had any news of him recently, but you know what his reply would be."

"A death wish?"

"Something like that. An urge for collective destruction, a relic of the distant past."

"A sort of archaism?"

"Who knows?"

"If Hitler had had the bomb," the girl says, "he wouldn't have hesitated to use it."

"'That's what people have often said to console me. It's also what I sometimes tell myself. Come and see.'"

HE LEADS HER over to one of the doors, which he opens. Beyond it is an unpretentious room in a seaside house—on Long Island, he says. The waves can be seen and heard, the air is fresh despite the heat. The sky above the Atlantic is sprinkled with small white clouds.

There are two men in the room, their foreheads bathed in sweat. Einstein, who says that it's July 1939, introduces them as Eugene Wigner and Leo Szilard, two physicists whom he trusts implicitly. He has known the latter for a long time. A Jew like himself, Szilard left Germany in a hurry with his savings hidden in his shoes.

Einstein quickly explains the situation to the girl, whom his fellow physicists show no sign of noticing. They have come to persuade him to write a letter to President Roosevelt. In order to overtake Germany, Roosevelt must be urged to enter America in the race for atomic fission, nuclear energy and ultimately, no doubt, the bomb.

They have come to ask this of him, the staunch pacifist.

"I was in an exceptionally dramatic position. Can you imagine? There was a possibility that the fate of the planet was being decided in this little seaside house, which a doctor friend of mine had lent me. We knew that the fission of uranium had been achieved and that several teams were on the way to producing a chain reaction. The French, in particular, were working on it with Joliot-Curie."

"The Germans too?"

"We weren't sure, as I told you, but Niels Bohr thought so. He was extremely worried."

"Could you tell me a bit about him?"

"Now?"

"Or later."

"Oh, I was very fond of the man. People call me a masochist when I say that, because he was always badgering me, but it's true. Niels was a Danish physicist who headed the Copenhagen Institute. My toughest opponent. Well, now that we're on the subject, let's indulge in a digression . . ."

He goes over to another door and opens it. Seated in an armchair in the room beyond it is a thin man of about sixty with short hair and a short, narrow torso that suggests he's below average height. He seems wholly engrossed in a long daydream. Einstein greets him warmly and says a few words in German. Niels Bohr makes no reply. We can't tell if he has seen or heard a thing. An almost ghostly figure, a thinker from the museum of shades, he doesn't stir.

"He's often like this," Einstein says with a smile. "A bucktoothed dreamer."

"What is he dreaming about?"

"Arguments he can use against me."

"And you liked him?"

"I almost loved him. Especially at first, before he became dogmatic. He was like an extremely sensitive child living in a sort of trance. There are many stories told about him. For instance, he used to take up his position on the beach facing the sea and perform some calisthenics, then stretch out his arms and remain motionless for an hour or two, without a tremor, transfixed by his own thoughts. I can understand that."

For a moment he looks at Bohr with something akin to affection. Then he takes the girl's arm and leads her away. Come, his manner conveys, let's not disturb him.

They leave that door and return to the one that opens onto Long Island.

"He expressed himself badly most of the time," Einstein says, still on the subject of Bohr. "The sentences he uttered were confused—very long-winded, very muddled. He would break off and start again. 'Couldn't you make an effort to speak better so we can understand you better?' he was asked on one occasion. And do you know what he replied?"

"How could I?"

"He said, 'I try not to speak more clearly than I think.' I like that statement immensely, don't you?"

"Why did he become the head of a school?"

"I was told another story about him," Einstein says without answering his visitor's question. "One day he invited some science correspondents to his home to bring them up to date on his ideas and research. His son, a Nobel laureate like himself, was also present. Niels addressed the journalists for hours, bumbling along in his customary way, until darkness fell. He invited them to spend the night at his house or a nearby hotel, which they agreed to do. The next morning they all met for breakfast, after which Niels told his son, 'Please give us a brief résumé of what we said yesterday.' His son obediently complied, taking about fifteen minutes over it. Niels listened to him attentively. Then he said, 'Succinct, lucid . . . and wrong.'"

Einstein's explosion of mirth is even louder than usual. He almost weeps with mirth. Then, apparently remembering the girl's question, he stops laughing.

"What made him famous? Quantum mechanics, of course."

A sudden doubt seems to assail him.

"You know what that is?"

"More or less."

"All at once, in my youth, physicists began to study the infinitely small. It was the great novelty. Atoms, then particles. Protons and, later on, neutrons forming the nucleus of atoms. Electrons dancing around, photons and other things as well. We were tiptoeing into the hitherto unsuspected territories of the invisible. It was a breathtaking venture. We went from discovery to surprise and surprise to amazement. For the first time in history a whole world had revealed itself to us, and that world, which clearly displayed the emptiness of matter, was well and truly our own, the one in which we live, the one of which we ourselves are composed. It wasn't long before some of us observed irregularities and oddities—anomalies, we called them—in the behavior of the elementary particles we were discovering and christening by degrees. A little later on, those observations led Heisenberg to formulate his celebrated 'uncertainty principle.' On that level, it held that all matter—we ourselves, in other words—contains a proportion of probability, of the irreducible unknown."

"And you found that hard to accept?"

"So people have said, but they're wrong. I never disavowed quantum mechanics, whose birth I'd assisted in the quantic period. It has rendered us genuine, incomparable service and still does. It's at work in your wristwatch, in your electronic notebook, in countless everyday objects. I would go so far as to say that life would be inexplicable without it."

"Life? Real life?"

"Absolutely. Because, to put it in a nutshell, only quantum mechanics enables us to account for the stability of atoms' internal states. It is thanks to that stability, which classical mechanics never managed to elucidate, in other words, to solve the question why atoms remain what they are, that chemical structures can reproduce themselves and engender life."

"So life isn't classical?"

"That's one way of putting it. Neither is the world."

"What was it that put you at odds with Bohr?"

"Some physicists said it was a simple matter of aesthetics, others of metaphysics. Others say that each of us felt sure the other was mad, which is a classical symptom of all known cases of genuine insanity. In fact, despite or perhaps because of his terminological difficulties, Bohr possessed a greatly expanded conceptual field and an exceptional physicist's vocabulary that aroused my admiration. You would have thought he was an imaginative semanticist. Taking advantage of his elocutionary problems, he invented figures of speech and had an answer to everything. When we presented him with a major objection, he could spend whole nights cudgeling his brains in a sort of abstract confrontation with himself, obsessed with notions like complementarity or contradiction or fortuity. In the morning he would turn up looking bleary and tell me, 'Albert, I've got the answer.'"

"But wasn't it you who helped to give birth to the new physics, only to dispute it later?"

"I had an inkling of what would happen, it's true. I saw it coming and drew my own conclusions very quickly, but I never really disputed it. I debated it step by step, because I felt it wasn't wrong but incomplete, and Niels was my principal opponent. With the passing of the years, the Nobel Prize and personal success—all well-deserved, because we owe him some remarkable advances—he succumbed to our favorite defect: he became brusque and authoritarian. He believed he'd discovered everything within the space of a few years, so we wouldn't get much further. As a result, some people said that he'd been marking time—even that he was opposed to new breakthroughs. He seemed at ease in his quantic fog, and everything had to be as he said."

"That didn't apply to you?"

"No—at least, I hope not. The fact is, I've always been something of a waverer—an opportunist, you might say. I endeavored to cling to a few simple principles of clarity, comprehension and utility. I strove to uphold the idea that we engage in scientific research in order to fathom the secrets of this world, not to render them more obscure still. At Princeton during the 1930s, this often led people to think I was an old fool."

"Why?"

"Quite simply because my attitude seemed out of date. Once regarded as an adventurer, I was relegated to the rearguard. I was accused of clinging to my established views and lacking all understanding of modern concepts."

"There's been a rethink, I gather."

"I'm well aware of that. Contemporary physicists—some of them, at least—have rehabilitated me, so to speak. General relativity has conquered the stellar world; quasars, pulsars and black holes have come to my aid like the gravitational waves we're told about. All these newcomers have reinforced my theory, but the feeling that prevailed at the time was that theory, and theory alone, was capable of defining the scientist's sphere of activity. As if the world possessed no guaranteed reality beyond our own perception of it. As if that perception were essential to the existence of that same reality. As if, in the frantic effervescence of things observed on the level of the invisible—in the irregularity of particles, in the fact that one of those particles could be both here and there; in other words, that a door could be open and shut at the same time—all possibility of a universal determinist law were disappearing. As if the world were losing all consistency and chance were finally imposing itself on the human mind."

"Did everyone share that view?"

"Nearly everyone. The new dance was led by Bohr in Denmark and Heisenberg in Germany. What they said was this: We've reached the limit of the sensory. If we wish to proceed further, only the language of equations can accompany us there. We must abandon any idea of a sensory representation, of a habitual, traditional approach, of mental apprehension, and navigate by equations. Schrödinger was one of the few who stated that we could go beyond equations, and that other images awaited us there. I thought so too, or rather, I hoped so. I didn't care to stray into territory where we might suddenly be ambushed by new laws, which were even termed laws of uncertainty."

"What was it that upset you about these aberrations?"

"The others were luxuriating in probability and uncertainty. This led them to speak of oddities, 'delocalized electrons' and heaven alone knows what else. Personally, I couldn't and still can't accept unpredictability, fortuity, arbitrariness, fundamental disorder. Call me what you will—irrational, if you like. I said it again and again, I wrote to Niels and others. Probabilism wasn't enough for me. I felt that the route they'd taken, tempting though it seemed, afforded no way out. I felt they were leading us down an avenue that abounded in ingenious brainwaves, dizzy flights of fancy and attractions of all kinds, but which came to a dead end. There was something missing, some fundamental thing that still remained hidden."

"What do you hope for from the world?"

"I've never changed in that respect. I want nothing, in fact. I would simply like the world we've undertaken to know to be well ordered and harmonious. I'd like it to be accessible to us, so that our minds can plumb its innermost secrets. That's how I conceive of the world, or, more precisely, how I feel it is. Even today, after so many vicissitudes."

"And Niels Bohr didn't agree?"

"He wasn't alone!" Einstein replies with sudden vehemence. "There was a whole bunch of them! The Copenhagen school and others claimed that science had reached a sort of point of no return; that we must give up passing judgement on reality and resign ourselves to irremediable disorder, to the impossibility of knowing everything at the same time—the velocity and location of a particle, for instance!"

"Whereas you needed order of some kind?"

"Something like that. But I dislike the word 'order,' it's politically and socially infelicitous. I'd sooner speak of harmony, of complex but elegant solidity. Without that, I couldn't understand why we were devoting ourselves to science. And I still can't."

"When you were talking to Newton you criticized him for dragging in God the repairman, but didn't you yourself say that God doesn't play dice?"

"I may have phrased it like that for simplicity's sake. I

didn't put it exactly that way, either, but the collective memory abridges everything. I've never imagined a god, or just a demiurge who makes the universe work by pulling strings and cranking handles. To me that makes no sense. We've ceased to be like Descartes, who claimed to be searching for the laws God allegedly installed in nature as if he were fond of playing hide-and-seek—as if he were a school principal, or a sergeant-major, or a lawyer, or the organizer of a treasure hunt. I've referred to a mysterious flutist who plays somewhere far away, and whose music I can faintly hear. But the word 'flutist' mustn't be taken literally, of course. How stupid, how scandalous to turn God into a person! To do that is to reduce an immense mystery to the scale of our own pathetic little concerns—to insult the universe. No, what I mean is, I've always felt the need for an arrangement of things that would result in perfect concord."

"Concord between whom or what?"

"Let's be modest: between the universe and ourselves."

"And this concord could one day be embodied in a single equation?"

"A single, magnificent equation. The equation of the universe!"

Once again she pauses for a moment, silently contemplating this image—but is it really an image?—of a universal equation that contains everything and its opposite: a definitive, irrevocable formula.

"Is that why you go on working?" she asks at length.

"Can you think of a better reason?"

- SEVEN -

ON HER OWN initiative, the girl returns to the subject of nuclear weapons.

"Did Niels Bohr share your uneasiness about the use to which the bomb could be put?" she asks.

"And how! Indeed, I think he was more apprehensive than any of us—sometimes almost terrified by the idea that a single country could monopolize the power of death. Did you know that he went to see Churchill towards the end of the war and urged him to give our nuclear secrets to the Russians?"

"To Stalin?"

"Certainly! Churchill thought he was mad. He practically threw him out."

"Not that it prevented the Russians from getting their own bomb."

"No. Terror seeks an equilibrium too. Even at the end of the war, with victory won, Niels was still just as worried about leaving the bomb exclusively in American hands. He expressed his deep concern—he even predicted the advent of the hydrogen bomb and the development of unlimited destructive power. He dreamed of an agreement between the great powers aimed at limiting the use of the new energy."

Both men, one in the United States, the other in Europe, doubted that the military secret would be kept for long. Once the book of nature has been opened, Bohr used to say, anyone can read it. And spies are there to supply the keys to reading it—to the highest bidder. Spies, Einstein would say, who must have done some physics. That's something, at least.

How could a planetary catastrophe be avoided? That was the immediate question. Humanity wasn't ready to enter the nuclear age, Einstein thought. Everything had happened too suddenly because research had been accelerated by this accursed war, which had to be won at all costs. That being so, what? How should scientists behave now that they had suddenly, and contrary to all expectations, become modern sorcerer's apprentices and exterminating angels? How should they react? How should they respond to American and British politicians whose sole concern, at that stage of the war, was victory?

Was this the beginning of a new relationship between science and power? Now that physicists had become valuable and suspect, would they be escorted under guard to luxurious compounds, there to work on methods of annihilation?

Would they necessarily be subject to the powers that be? Would they acquiesce?

Einstein tells his visitor that all these questions—and

many others—were asked between 1939 and 1945, starting with the most obvious. Was an atomic weapon practicable? Should one be put in hand? Would we be able to keep it secret? Above all, if we manufactured one, could we—should we—use it?

The girl finds herself back in the little seaside house with Szilard and Wigner. Einstein has joined them. His hair is fluttering in the wind and he's holding a dead pipe in his fist. Szilard hands him a draft written on one side of a single sheet of paper. He reads it calmly, ponders for a while and asks some inaudible questions.

Szilard and Wigner reply in unison. They do their best to be convincing, become insistent. Both of them point out that Einstein grasped the danger very quickly. Yes, two German chemists, Otto Hahn and Fritz Strassmann, have manufactured barium by bombarding uranium with neutrons. There's no doubt about it: atomic fission is in progress. It's not only possible, it has been achieved! Another German physicist, Harteck, has just informed the war ministry that this discovery opens up the prospect—what a dismal military aspiration!—of producing an explosive superior in power to conventional bombs "by several orders of magnitude." Minister Goebbels is said to be delighted. Hahn, one of the two chemists and the person who tried to warn Niels Bohr, feels like throwing his barium into the sea and committing suicide at once. The Germans, who have just annexed Czechoslovakia, have imposed a total ban on exports of uranium, which seems to indicate that they mean to make use of it.

Where can more uranium be found? In the Belgian Congo—there above all. Elisabeth, dowager queen of Belgium, knows Einstein quite well. They regularly exchange cordial letters and even make music together. Couldn't he quickly warn her, put her on her guard?

Einstein thinks a while longer. No, he won't write to the Belgian queen mother. Yes, he'll write to Roosevelt to put him in the picture and inform him that humanity is setting out on a new road. The letter, probably drafted by Szilard, is signed by Einstein on August 2, 1939. A brief missive, it states that recent scientific advances have enabled uranium to be transformed into "an important new source of energy," and that a chain reaction is under consideration. This chain reaction would make it possible to manufacture a bomb, which, if deployed against a seaport, for example, "might very well destroy the whole port together with some of the surrounding territory."

It was not until several months after the letter was sent (Poland had been invaded and war declared in Europe) that the Manhattan Project was officially launched at Los Alamos.

On July 16, 1944. the first "atomic bomb" was detonated in the New Mexico desert. On that occasion the director of the program, J. Robert Oppenheimer, is said to have quoted from the *Bhagavadgita*: "I am become death, the destroyer of worlds." Having come down to earth to restore peace and preserve the essence of life, which was threatened by a vast human war, Krishna had no choice but to put the fatal equation into effect. Not long afterwards

Oppenheimer would say, in a phrase with equally religious connotations, that science had experienced sin. It was an admission.

Einstein leaves his two friends in the house beside the sea and rejoins the girl. He closes the door behind him, still looking pensive, and brushes some grains of sand from his hair.

"And then came Hiroshima?" the girl says.

"Yes, the following year. The bomb prepared for the Germans was eventually dropped on the Japanese. Believe it or not, I first heard the news on the radio like everyone else."

"You took it badly?"

"It seems I called it a calamity."

He remains silent for a moment as if remembering that day, the first day in the new history of the world. It's a memory that doubtless haunts him often, even in his present state: the memory of the curse he bears. He will always go down in history as the father of the ultimate weapon of annihilation—he, who dreamed of peace, concord and international amity.

Without his flash of inspiration in 1905, without his sudden vision of the energy latent in matter, would Hiroshima have been destroyed? Human history follows overgrown paths and treads them only once. It's impossible to retrace our steps through such a jungle of causes, fortuities and effects.

Another brain would doubtless have had the same idea a little later, we can be almost certain of that. Long signposted on the road of discovery, the violent suicide of matter was patiently awaiting its appointed hour. But the road that led from Bern to Hiroshima went by way of defeated Germany, the Versailles Treaty, the Wall Street Crash, the persecution of the Jews, Einstein's exile to the United States, Pearl Harbor, Roosevelt's death and Truman's executive decision. How to ensure that, among other historical scenarios, the city of Hiroshima would be the final target?

WE CAN ALSO ASK ourselves this: Should responsibility for any particular action, real or fictitious, endure to the end of time? Do we continue to labor under our sins and crimes throughout the ages, even when no intention to inflict harm can be attributed to us?

Once upon a time the Church used to exhume heretics' corpses. Richly attired bishops armed with silver crosses abused their bones and spat on them. Their dust was officially anathematized and often committed to the flames. We find such customs surprising, but they were still current in Europe at the beginning of the eighteenth century—only the day before yesterday—when old King Louis XIV, kneeling before Madame de Maintenon, ordained similar punishments, long after their death, for the exhumed remains of former residents of the Abbey of Port-Royal.

It was as if somewhere in the earth or the air there remained some ancient sin, some poison more tenacious than life itself.

The girl could ask the same questions in regard to

Einstein, with whom she has been for . . . How long, in fact? An hour, two hours? She can't tell. The scenes beyond the doors of his study appear sometimes by day, sometimes at night. There's no window through which the street can be seen.

Was he really the initiator and trailblazer of the route to the bomb and the nuclear submarines that now lurk on the ocean bed? Is he regarded as such by the mysterious authorities who allotted him his present abode? Is it a rather refined, even sophisticated version of the Gehenna of old, where guilty souls vainly wailed and groaned for all eternity?

Has the girl mistakenly rung the bell of one of the secret doors to hell?

Anything being conceivable, that too is possible. He's here, the new Sisyphus, doomed to undertake a tireless repetition of his calculations, which are forever erased and rewritten. But who put him here? Who passed judgement on him? Who keeps watch on him? Is this simply, as in the Buddhist samsara, an established state not subject to any decision-maker, law or master? Is it an exact image of the real way of the world, not that the world itself seems to realize this?

The girl reflects that she could entitle her piece *Einstein's Punishment*—a sort of hellish work of fiction—but she quickly dismisses the idea. Having just touched on Buddhism, she prefers the gentle image of a bodhisattva, one of those rare and superior individuals who, having finally earned the right to terminate their series of reincarnations and enter nirvana forever, choose to remain among mortals still in torment and help them to emerge from that cycle.

Einstein the Bodhisattva. Yes, she likes that better.

SHE LOOKS AT HIM.

His shoulders are drooping. He's stroking his chin with three fingers, as he often does when thinking. All the study doors are shut. The room is at an even temperature. The girl has yet more questions to ask, but perhaps she tells herself that it's time to go. She has seen and heard enough. Whatever the real condition of this semblance of a man, she probably ought to leave him alone with his fatigue and lassitude—and with the ever faithful Helen Dukas, who must, unless her hours of work are over, be waiting in another room somewhere.

"Well . . ." the girl says, holding out her hand.

"No, wait, that gives me an idea! Don't go!"

Einstein's eyes have brightened. He squares his shoulders. Ignoring his visitor's outstretched hand, he opens the door to the waiting room.

"Isaac?" he calls. "Can you spare a moment?"

Newton appears a few seconds later, clutching a bunch of papers. He starts to say something, still looking sulky, but Einstein cuts him short.

"Just tell me this: are you convinced?"

"Not really," Newton replies.

"I thought as much. Listen, ditch those papers, I've got something much better. I have a proof to show you."

He tugs Newton towards one of the doors.

"A proof of what? What proof?"

"The proof of one of those equations you dislike. Proof that matter and energy are the same thing. I'd sooner not have had to show it to you, believe me."

"What do you mean?"

"Come this way."

"Where are you taking me?"

"Just to that door. There's something out there you must see, and I'm sure you'll understand right away. Don't be nervous, just look . . . See what they've made of our work . . ."

He opens the door. Instantly, a mighty wind seems to blow through the study, ruffling Einstein's hair and Newton's wig, snatching up countless sheets of paper and hurling them across the room. The girl clings to a piece of furniture. Newton's portrait becomes dislodged and crashes to the floor. The blackboard topples over, spilling its sticks of chalk and snapping them.

On the far side of the open door they see a nuclear explosion and its aftereffects: vaporized buildings, a devastated city, human shadows imprinted on walls.

Newton, who is standing behind Einstein, stares aghast at the destruction of Hiroshima. His wig flutters violently in the shock wave, which carries it away. He stares at the scene with bated breath.

"God Almighty," he mutters faintly.

THE THREE OF THEM stand there in a daze when Einstein shuts the door and they're back inside the shambles of a study. Newton looks around wildly, opens his mouth to speak but fails. He turns on the spot, trampling scattered papers and fragments of chalk underfoot, then stoops to retrieve his wig. Straightening up, he staggers a little as if unable to keep on his feet, as if his strength is deserting him.

The light in his eyes has dimmed. Something seems to be going on inside his body, which is still swathed in the black cloak. It's as if he's slowly dissolving, as if his flesh is losing some of its consistency, its solidity, the indefinable substance of which it has been composed heretofore. His breathing is almost inaudible.

"Are you all right?" Einstein asks him.

No response. He probably hasn't heard. He raises one of his hands to his eyes and looks at it. Can he see it? Impossible to tell. The hand turns white, very white, then translucent and finally almost transparent, all within a few seconds. He puts it to his eyes, intending to rub them, but his eyes, too, have undergone a change. They've turned pale, almost white, like a pair of snowflakes. Einstein and the girl can actually see them through his hand. They can also discern objects in the study—the overturned blackboard, a closed door—through his body.

Does Newton realize what is happening? Is he still conscious?

"Goodbye, Isaac," Einstein says.

His reprieve expired at Hiroshima: atomic fission is carrying him off. Quite motionless now, Isaac Newton is disappearing forever like his clothes, his wig, his buckled shoes. He's just an evanescent silhouette that gradually merges with the air until there's nothing left of him.

- EIGHT -

So many questions remain.

For instance, people have often written that the last hurdle to be cleared before attaining the Great All is how to reconcile two theories that now share the universe between them. Both are justified, effective and elegant. They are gravitation (in other words, general relativity) and quantum mechanics. Each describes and explains the world in its own way and on its own level, but they cannot function together. They are mutually exclusive. One cannot admit the other to its domain and vice versa.

When questioned about this by the girl, Einstein isn't very talkative. Yes, of course he's familiar with the problem. Like other people, he wrestled with it for a long time. When he taxed Bohr with deficiencies in quantum theory, that didn't mean he discarded it like a dirty rag—anything but. He himself had been one of its initiators. He acknowledged its acrobatic virtues, its subtle analysis of the impalpable, the finesse with which it had laid down new laws.

But he also perceived its failings, its lacunae, and that danger to science itself which he always sensed in the notion of uncertainty or, to put it more simply, of chance.

However, since he's careful to keep himself up to date (as witness the mail he receives), he now knows that other theories are gaining ground. Foremost among them is that of strings and superstrings, a vision in eleven dimensions, not four: one dimension of time and ten of space (tough work for the mind, but it's got to get used to the idea). This is a vision that delineates another, invisible, world with the aid of imaginary but pertinent entities like strings, membranes and branes; a world constituted like a virtual landscape, plaited, rocking, moving, crumpled, entwined, undulating, sometimes open, sometimes closed, and sometimes all of these at once.

Some scientific institutes assert that this reality will be that of the century we have just entered. We are going to change realities, change worlds. The two contradictory theories will both be swallowed up by the complex fluctuations of the new vacuum in which they will become reconciled.

What is a vacuum? Strange question. Is there any definition of empty other than "the contrary of full"? Certainly. If physicists are to be believed, a vacuum is a system's minimal state of energy. Yes, but there are several systems, so there are several vacua. As many vacua as systems. That's how it is.

For we are definitely dealing with a vacuum, one that isn't nothingness but vibrates with a latent energy that was estimated to be immense but now, according to our latest information, appears to be incredibly weak: barely conceivable, barely quantifiable—infinitesimal but not inexistent.

Why is it so weak? By chance, so some people seem to say: because the history of the world has developed in that manner, in accordance with a largely fortuitous series of circumstances. But Einstein—and he is not alone—persists in believing and hoping that a strict law has presided over this constituent weakness of things, and that we shall someday be able to define that law at last.

Will it be the very last law of all, the one that closes the great book?

Einstein has no quarrel with vacuum energy and string theory. As he has often said, he is quite prepared to abandon his views if they're proved to be untenable. He has already done so in respect of the universe itself, which he judged to be static but is not—in fact, its expansion is accelerating. No one can accuse him of clinging to defunct theories or of being intellectually unadventurous and uninventive. Barely a hundred years ago he stated—and proved—that space and time do not exist in absolute terms. He replaced them with space-time, which has largely demonstrated its convenience and is itself an absolute like the speed of light. And now space-time has been extended and even enriched in its turn, for string theory, too, is jettisoning obsolete concepts as it proceeds, confirming that each force corresponds to a curved dimension, multiplying spatial dimensions and abolishing punctual particles.

A triumph of matter. A triumph, too, in the still inaccessible, impalpable heart of the invisible.

"What's your opinion of these things, Mr. Einstein?"

He doesn't have one. If the two theories that share the world are reconciled by superstrings, so much the better. It has already happened with light, which is both wave and particle. That idea seemed unacceptable, yet electric lights still worked. After all, science must be allowed to describe the universe even if the universe is indescribable. We must know how to lay aside logic, which is just a banal mode of speech, a human convention to which things are not necessarily subject.

A simple question of terminology, perhaps. In any case, you can't comment on what is plain to see.

SOMEONE ELSE WAS awaited on the great stage of the invisible.

Towards the end of the 1960s, by dint of purely theoretical activity, a Scottish physicist named Peter Higgs invented an unknown material particle termed the "Higgs boson," which no one has ever observed. The universe had acquired yet another phantom, a particle relatively heavy for its minuscule size. If some trace of this is discovered (possibly in 2007, by CERN's new particle collider near Geneva), some believe that it will present us with the supreme explanation of the union of all the sometimes opposing forces that can still be distinguished. This would crown what physicists call the Standard Model and secure harmony and coherence in the realm of particles.

The Higgs boson? Yes, of course Einstein has heard of

it, and he's waiting impatiently, like everyone else, for it to manifest itself. There are days when he even wonders if he's being kept on ice for that revelation—at least until 2007.

But other people assert that the Higgs boson isn't the key to the universe after all. They say it must be sought beyond the Standard Model and more in the direction of "Susy," their pet name for supersymmetry, a mathematical operation, the mother of all particles and all energies and the indispensable stepping stone to superstrings. The Higgs boson and Susy . . . They might almost be passwords.

"Forgive me for asking," the girl says, "but what's the Standard Model?"

"It's the accompaniment to quantum mechanics' dream of unifying two of the four forces, electromagnetic and weak nuclear. And, before long, strong nuclear. It's the beginning of the long march."

"And Susy?"

"Susy would be a big step towards the hoped-for unification of the Standard Model and gravitation. Ergo, of four forces. It would unify the ultra-sociable bosons and the antisocial fermions, which we've already mentioned, and which have hitherto been neatly parked in watertight compartments, because it posits the existence of bosonic counterparts of fermions and vice versa. Nothing is irreducibly discrete, so there's hope yet."

Einstein adds that the celebrated superstrings would be Grail, Moby Dick, Simurgh, Promised Land and nirvana all rolled up into one, and that they seem quite suited to effecting the ultimate unification of the four forces. Only conceptually, of course. Only theoretically. There are no practical applications at present. Experiments designed to reveal Susy are in hand, but none for superstrings. Triumphing by means of hypothesis and calculation, that's all we can hope for. We can't expect matter not to be what it is, but we can dream of it sometimes.

At other times Einstein almost wistfully reverts to himself, reflecting that certain theories, like this one, have all the appearance of a dream, a physicist's utopia where the magic of the invisible comes into force in reality.

But wasn't he himself, when envisioning curved space or moving the stars in his mind at the age of twenty-five or twenty-six, immersed in a dream that a persevering Englishman, assisted by a solar eclipse, was destined to bring true?

THERE REMAINS ANOTHER lingering question bound up with the previous one but more important by far. It relates to the oneness or multiplicity of the universe. The physicists who call themselves realists, who take the equations of quantum mechanics seriously, who espouse pure, hard theories that are often confirmed by experiment, who dispense with common sense and everyday life and do, in effect, leave it to science to define the world in cold, objective terms believe that this universe of ours is multiple. According to them, we are only one bundle of reality in a vast number of others.

To physicists of cosmological persuasion, who are enthusiastic advocates of universal inflation and the quantic vacuum, we live in an explosion of champagne, a continuously renewed effervescence. It's impossible to disprove this state of affairs either way. That's how it is, we're told; theory indicates as much. Universes exist even if we don't observe them, and we have to accept the fact.

Even though the dimensions of the universe we can observe seem gigantic to us (twenty-eight billion light-years in apparent diameter), it is only one among other possibly receding and long inaccessible universes that are concealed from us. Let us call the physicists who espouse this theory "realists" or "neo-realists."

Opposing them are the neo-idealists, who also (strangely) call themselves "positivists." In their view, affirmation is impossible without knowledge, and nothing really exists or merits the epithet "existent" unless we can observe it. A "multiverse," or universe unobservable by definition like the multitude of other universes postulated by the realists, has no real existence. To exist in this sense is to be with others, to form a relationship, to be perceived. Scientific existence, even existence pure and simple, entails a close alliance between the observer and the observed, each acting upon the other—a very old view, formerly advanced by Asiatic visionaries, which culminated in quantic equations.

Why shouldn't one speak of a multiplicity of universes? Many poets have done so. But doesn't that smack more of psychiatry, of an imagination given over to itself? Not at all, the "realists" retort. The brain in need of treatment is yours, which claims that science can't exist without it. It's one form of madness versus others.

"I'm waiting," Einstein says. "I told you: there'll come a time when my ideas, which were unacceptable in their day, will be secondhand goods. Unified field theory, the great white whale on which I pinned so many hopes, will wind up in the flea market."

"But tell me," says the girl, who has decided to stay a little while longer, "I'm still under twenty-five and doing my best to follow the development of ideas in my own day. I want to know what world I'm living in. I may have children . . ."

"That's quite natural."

"And I sometimes seem to hear you say, you and the others, that macroscopic objects like us exist—objects like me, anyway, and horses and planets and stars—but that these observable objects, the ones we rub shoulders with in our 'bundle,' consist of infinitesimal entities that *don't* exist. Not in the local sense of the term, not here and now."

"Yes," he says with a weary smile, "that's it, more or less."

"It's far-fetched, though, you must admit."

"I do admit it. Besides, I've always maintained, and still do, that this is where the shoe pinches. It's where our minds stop short, unable to go any further."

"But both assertions are true?"

'They aren't assertions, they're observations. A person who makes an assertion will always find someone to contradict him. It's harder when someone makes an observation. His contradictor has to make an observation of his own. As in this case, incidentally."

"So?"

"I think it all comes down to what one calls the truth. I'm not sure if it's easy to define that word. When I was studying philosophy, one standard subject for essays was: Is truth the opposite of a lie, or the opposite of a mistake?"

"What was your answer?"

"If some of my colleagues are to be believed," Einstein replies with a final laugh, "I said that the opposite of truth is truth."

After a brief silence, the girl admits to feeling rather at sea.

"That makes two of us," says Einstein. "I'm waiting. Towards the end of my life, having heard it said so often, I acknowledged having been mistaken, but today I'm not so sure. It's strange. Sometimes—only sometimes—I now understand why I persisted in stating with an obstinacy people called senile—what would they say now, assuming I still have an age!—that physics must represent a reality, a reality in space and time, in space-time, because we can't conceive of any other. If I'm still being allowed to work, I tell myself with a trace of optimism, it's because I wasn't entirely wrong, because I'm not chasing shadows and there's something still in store for me. If physics gives up representing a reality, what's the good of it?"

They revert to the non-separability concept dear to the upholders of quantum mechanics: no existence unless that existence is perceived and no perception devoid of an effect on the percept. It undoubtedly works on a certain level, Einstein says again. It's true, it can be ascertained. But this would mean that everything is quantic, even the organization of our bodies, even the cosmos, when all our senses and all our powers of reasoning tell us the opposite. Things are discrete everywhere: as in geology, biology and astrophysics, which are also sciences, so in daily life. They're separated by space and time, or space-time, if one prefers. Otherwise there would be no events, no relationships, no history. That's how we study things—how we're able to study them. If gravity is to act on two bodies, those bodies must be discrete. If not, what are we talking about?

The girl now alludes to the old EPR paradox to which Einstein lent his initial (the others being those of Podolsky and Rosen), the crucial—more recent—experiment conducted by the physicist Aspect and the research carried out by other experts of whom she has heard. What about these particles that receive information instantaneously, wherever they happen to be in the universe, as if space and time had no hold over them. As if they ignored and dominated them—or constituted them?—and as if non-localized influences faster than light were at work? What about them?

Einstein shrugs his shoulders, takes a few steps, fidgets, shakes his head. A long time ago, he says, he and two other physicists (yes, in their celebrated EPR paradox) represented that unthinkable thing, that unreality, as a logical consequence of quantum mechanics. They called it a paradox to show that it was unthinkable. Could science be paradoxical? Could it run counter to physical sensation and mental logic?

"What if the problem was badly formulated?" he asks suddenly.

"Badly formulated by whom?"

"What if it was just a matter of words? What if space-time on the one hand and the conservation of energy on the other were incompatible, as Niels Bohr contended, and were really just words? Listen, I honestly can't say why and I'm more and more doubtful of having time enough to find a satisfactory answer, but I can't give up. You understand? I can't *refuse to explain*. There it is. I can't bring myself to admit defeat and say: Beyond a certain point the world is genuinely inexplicable, prodigiously incoherent and fundamentally paradoxical, and I will never know how or why. I will never be able to say that. You asked me the question on arrival, remember? You asked me to explain, and I told you that explaining is the hardest thing in the world. Now do you see why? Because I'd have to explain that we must give up explaining. And I never would! It would mean going against all that made up my life. Was I lionized, feted, decorated, celebrated, showered with awards and praised to the skies, only to take my leave, sticking out my tongue for the last time and saying: Ladies and gentlemen, I've been no earthly use, I've floundered around in ignorance, I don't know what to say to you and I'm making for the exit bereft of ideas?"

He sits down and puts his head in his hands.

Stuck in space-time (she doesn't know how or when she'll be able to extricate herself), the girl catches on at last. She has reached the nub of the problem. It concerns the end of his relationship with the world—indeed, the very existence of that world. According to the fanatical skeptics on the one hand and the champions of illusion on the other, that existence is not a foregone conclusion. The stone I trip over and cut my knee on when I fall isn't necessarily what it appears to be. It may be another stone in another version of the cosmos, or a piece of wood, or a cork, or an anthill. It may also be none of those things—in fact it may not "be" at all. In that case, why do I feel pain? Answer: What pain?

Einstein isn't in that camp, he has already said so. Very few scientists will defend a generalized illusion, like a backdrop, except for fun or in provocation. Those who study things don't like to envisage the possibility that they're working on phantoms.

They are on the side of Parmenides: that which exists, exists, and that which doesn't exist, doesn't exist. Nothing else has any existence.

What the girl also rejects is the so-called anthropic notion that the world was created, or put in place, with the intention that it should ultimately become known to us—an idea of such half-baked arrogance that it seems to be quashed by ridicule even before being formulated. Unless, of course, we fall back on the previous hypothesis that the world doesn't exist, and that the meticulously detailed tissue of illusions surrounding us was woven solely to deceive us. But to what end and by whom? By malicious demiurges whose puppets we are, and whose sole aim is to amuse themselves? If so, who created those demiurges and gave them a liking for toys? Who takes care of all this? Why has this huge marionette factory taken millions of years to set up? We have no idea.

Just for fun, the girl speaks of a puddle formed in a pot-holed road after a shower of rain. Suddenly endowed with reason, the puddle explores the ground around it and cries, "What a miraculous coincidence! My shape and dimensions exactly match those of this hole in the road! That means I was meant to be at this particular spot on this particular road! There's no doubt about it, so what other purpose could I serve?"

Some minds, she says, are shaped like puddles.

Let's be content, she goes on in the easy, unconstrained manner characteristic of her ever since her arrival on the scene—let's be content with the less mythological, less theatrical idea that the world is within our reach; that it exists in some way or another, or at all events that "what exists, exists" (as Parmenides said); that this world is written in mathematical language and that we can not only dissect it and extract constants and laws but, by theoretical means, perceive its composition, evolution and—some billions of years hence—dissolution.

Reverting to a point she raised earlier, she says she has noticed that many scientists are surprised—greatly surprised, in fact—that the world corresponds to our calculations and obeys our brain. Some of them (including

Einstein?) even admit to being worried by this and fail to understand how we can understand what surrounds us.

Look at ants, says the girl, thinking aloud (and vaulting over her imaginary puddle). They're insects that live in perfect societies and are even said to resemble the individual neurons in a collective "formic" brain (so they swiftly exchange expected or unexpected items of information that modify their behavior). It never occurs to us to think that they believe the world to have been created for them, that it meets their requirements or that they dominate it. We even think they don't think at all.

If we imagine intellects infinitely greater, more complex, imaginative and flexible than our own, the girl goes on, intellects even more remote from ours than we are from ants, those intellects could never tell themselves, assuming they needed to do so, that the world had been created in order to be understood by us humans.

Just as the ant and the elephant are of the same size *sub specie immensitatis*, an immense intellect would detect no difference between our understanding of the world and that of ants. Brain for brain, we would be lumped together. Even though, when studied as closely as we study ants, human beings have perfected some rudimentary techniques and produced some works of art which they alone admire, they are, compared to these putative oceans of knowledge that will always elude us, no more than a feeble, endangered, recent subspecies, surprisingly limited in terms of space and time and afflicted with an egocentricity so presumptuous that it only highlights their ignorance, conceit and weakness.

The thinking of this species—infinitesimal relative to the universe—is necessarily restricted to itself. Ants think on the formic level and humans on the human level. Having established their own criteria, their own rules of verification, they are surprised when what they observe corresponds to what they have decreed.

Human thought, the girl goes on, is the only reference human thought possesses. As for the laws it discovers and verifies in the universe, there is no proof that they exist in an absolute reality, or that, even if they do exist outside us, they are valid. Being only a projection of ourselves, they are valid only from our own point of view. As the advocates of multiplicity, the realists, so insistently proclaim, it is even probable that the universe is not the one we observe and analyze, or at least that it isn't that alone.

Any other theory—other mathematical progressions and other experiments in verification in another version of the universe—would doubtless lure us onto treacherous ground and lose us there.

Every thought creates its own prison and escapes in its own way—or thinks it does, because we alert our guards at the same time as we dig our tunnels. The voice that asks the questions supplies the answers. Nothing ever transcends the human, neither religious speculation, which is stalled by dogma from the outset, nor daring philosophical forays; neither the extreme virtuosity (as we see it) of technical verifications nor our illusory breakouts. And for a very good reason. Our proofs are valid for us alone. They probably interest no one else except as local curiosities. Our sensory system, our logical thought and imagination cannot escape our own orbit.

During the 1990s some neurobiologists defined the human brain as "the most complex object in the universe." How did they know? The arrogance of that phrase does not disguise the fact that the brain defines itself and sings its own praises without even realizing that it is simultaneously criticizing and devaluing itself because it is capable of vanity.

The girl wonders aloud where this abiding presumption comes from. Perhaps, yet again, it stems from that old belief dating from a time when we thought the earth was big and situated in the center of the universe: the belief that man, the supreme miracle of Creation, was made by God in his own image—by God, who was only a superman himself—and that his mental processes, which exalt him above all living creatures, are essentially divine and incomparable. Don't scientists suffer from a vestige of that very ancient self-delusion, even today?

"It's quite possible," Einstein tells the girl. "Men of science are men like any others."

"Yet another form of reasoning that bites its own tail?"

"Yet another," he says. "Perhaps. It thinks it's a straight line and it's only a circle."

How many people have there been in the course of history who began by forming conclusions and called them principles? The list would surely be a long one. Bossuet thought he had floored the Protestants by 'demonstrating' that their affirmations were at odds with Holy Writ, which was just what the Protestants vehemently invoked. To both parties, these scriptures—which no one now doubts were of human authorship—constituted a crystalline, immutable truth, revealed for ever and imposing its light upon the world. The only weak point: as time went by, that truth disintegrated like a scrap of threadbare cloth being torn apart by several hands. It no longer meant the same to all the readers who invoked it, thereby proving that it wasn't the truth.

Einstein has been listening to all this with a smile, occasionally shrugging as if to say: Yes, yes, I know, so I've been told, so I've often been told, but for all that . . .

For all that, the temptation is a strong one. Stronger for him, no doubt, than for anyone. To have gone so far in his cerebral exploration of the universe, to have formulated so many unexpected but prolific equations in the maelstrom of his mind, to have sensed that the solution was quite close, within reach of his neurons, and to have benefited from an unwonted reprieve, only to put his head in his hands and say: I don't know which way to go . . . I thought I was right and I was wrong. I thought I was wrong and I was right. But does any importance attach to the fact of being right and the fact of being wrong? Do those words possess any meaning? Shouldn't those two concepts—true and false, this and that—themselves be consigned to the great human depository?

The girl guesses what's troubling him. "This definitive and universal equation of yours," she says, "what will you do when you've formulated it? Will you stay here and watch the flies on the ceiling?"

"There aren't any flies here, worse luck. I often regret that, because they're creatures that can fly, walk upside down and see the world through compound eyes. All the things we can't do."

"But flies lack the enjoyment of thinking."

"Don't poke fun at them. They know all they ought to know, which is far from being so in our case. As for enjoyment, they may well have pleasures that differ from ours. But they also have their limitations like us. Limitations that likewise differ from ours. For instance, I know that I don't know. They don't know that they know."

He adds that like himself—like a rhinoceros, like the sea, like the distant stars—flies are forever obedient to the mysterious flutist he mentioned, the one whose music he would dearly like to decipher.

He looks up for a moment as if in search of a fly, but there's none to be seen, no other visible form of life.

"What will I do?" he asks, still staring at the ceiling. "I don't ask myself that question. They probably won't keep me here once my work is done. Anyway, what would I look for? What would I say? Total knowledge implies the end of language. A zero language with nothing more to say."

"Where will you go?"

"Oh, somewhere in this floating universe. I will disappear from view, if there still are eyes to see. I shall probably go to the place we call 'nowhere.' I often repeat that word when I'm alone. Nowhere . . . It's a mysterious word, as strange in terms of space as 'always' and 'never' are in terms of time. A place devoid of location. What a surprising concept . . ."

"Will time still exist in this nowhere of yours?" she asks.

"Neither time nor space-time, I'm very much afraid."

"They're probably just figures of speech."

"How can one tell before one's consigned there?"

The girl asks if this perfect theory, this theory of the universe, this theory of everything (apart from nowhere), will be unassailable.

"Of course," he says, "or it wouldn't be considered perfect. It will have to pass numerous tests before it's posited."

"Will that spell the end of research, at least in physics?"

"No, no," Einstein replies, "on the contrary. When the equation of everything is placed on the pedestal it deserves and we're all paying homage to it, there'll be more work to do than ever. All our analyses and calculations will need redoing. Other doors will open, other dimensions, other territories. Only the adepts of ignorance consider themselves satisfied once and for all. When one knows nothing, it's forever. All knowledge leads to further obscurities, it's a well-known fact."

"Even the knowledge of everything?"

"That above all," he says.

"Why?"

"Because first one would have to agree on the meaning of the word 'everything.'"

"So what does it mean?"

"I have no idea."

- NINE -

THE GIRL TURNS OFF her tape recorder, stows it away in her bag and prepares to take her leave. Mechanically, she glances at her watch. It still isn't working.

Einstein, who regrets having had to simplify the scientific aspects of their discussion, asks if she's satisfied with her visit. It has exceeded her expectations, she tells him politely.

"What about you?" she asks.

"Me?"

A trifle surprised by the question, he mulls it over for a moment or two.

"If only I've managed to give you the beginnings of a taste for these things . . . A desire to venture two or three steps into those limitless expanses . . ."

"I had a taste for them already, that's why I came."

"I sensed that, which is why I agreed to see you."

"Thanks again."

"Of course," he says as he escorts her to the door, shuffling a little but still not making a sound, "you can always live like the vast majority and content yourself with slogans. You can also opt for ignorance. It's a choice that must be accepted, if not respected, because ignorance is reassuring. It makes an excellent umbrella. Three or four phrases will tell you all you need to know in order to live in this world for a while. Or rather, to live and die there and miss it altogether."

"The world isn't a simple place?"

"Far from it, that's why it surprises and perturbs people. That's why the timid protect themselves from it. As for us, who have devoted all our time to studying it a little more closely than others, when you chance to consult us, all we have to offer you is our own confusion, our questionable probabilities, our sometimes contradictory assertions. And you send us packing. It's hardly surprising."

"I haven't sent you packing, as you put it."

"But what a garden of delights it is! What an enchanted journey! If only you knew! What mental stupefaction and rapture, what daydreams!"

He's now speaking as if she has already left the room.

"The far-off and the elsewhere are within us. Who would have thought it? Who would have believed that such a journey was possible, that it would be interspersed with spells of vertigo, even of ecstasy—that it would compel us to question things and our view of them? Many of us lacked sufficient daring. They hesitated to set out on the great journey and drew back, closing their eyes and minds. I sometimes tell myself that I may be one of them, that I'm self-satisfied like Newton and have settled down in terrain

I've only just explored. But in that case, wouldn't I have been forgotten? Why would I have been put to work again?"

"Please tell me something before I go. What's on the physicist's agenda? What are the very latest things?"

"I'll list them for you very briefly, so pay attention. First, dark energy. What exactly is that repellent substance? And dark matter? And the shape and size of the supplementary dimensions? And that nagging question that confronts everyone: Why is the universe like it is? Why is it uniform? Why does it obey linear geometry? What formed the big structures, the galaxies? Was it inflation, that violent explosion at the very inception of our history? Why have we found more matter than antimatter? And zero time? And the Higgs boson? And Susy? What new routes, now veiled in mystery, are they going to open up for us?"

The girl flaps her hand as if to say: Don't overload me. He falls silent, though there's more he would like to say.

At the door the girl suddenly remarks that science, at least, is founded on optimism. It boldly presses on, believing in a future which, if not better, will at least be more enlightened. It is, perhaps, the only human activity that still refers to our future, to the future of knowledge and even of life. For science, no backward step is possible. It has to stride on ahead, sure of being better tomorrow and better still the day after.

To her, the word "progress" still means something. What does he think of it?

He doesn't think of it anymore. A kind of progress exists, well and good. Science has progressed since the troglodytic age, though one can't say the same for painting, still less for our morals. But can progress in our knowledge of the world and the perfection of techniques, lethal techniques included, be enough in itself? Is knowledge our sole ambition? Knowledge alone?

Somewhere, the girl recalls, Einstein wrote that the scientist's true happiness is comparable to that of the city dweller who, having left behind the turmoil of the suburbs and their disagreeable din, makes his way slowly up into the serenity of the mountains. From there his gaze travels afar through the pure, silent air and comes to rest on peaceful shapes that seem to have been placed there for all eternity. Does he still feel like that?

Yes, he says, sometimes. Only sometimes, although he no longer frequents the mountains. He's well and truly rid of the problems of the daily round, of illness, of persecution, of the rancors of family life, of press photographers. He's still troubled by embarrassing flashbacks—even, on occasion, by remorse. As for the shapes he contemplates, the images he sometimes sees when half opening his three doors, are they there for all eternity? He strongly doubts it, in fact he's sure the opposite applies.

Before the girl can make her exit he detains her once more.

"I know a new theory about the Big Bang that ought to appeal to you."

"What's that?'"

"You're aware that string theory refers to 'branes,' multidimensional objects that evolve like jellyfish on levels our perception cannot reach. Well, two of these membranes are said to have collided long ago, one material, the other possibly immaterial, and we are the product of that collision."

She stares at him in silence for a moment, doubtless trying to picture that primordial encounter. Then she asks, "What are you going to do now?"

"I'll probably play the violin for a bit after you've gone. I don't drink anymore, nor do I eat, sleep or smoke, but I've hung onto my music. It does me a lot of good. Then I'll go back to work."

"On the ultimate equation?"

"Perhaps, I don't know yet . . . I'm like a haunted castle pervaded by every passing wind. I sometimes follow the same mental route a hundred times a day. I retrace my steps, change a detail and start again . . . It's an everlasting muddle. The world is our muddle. And yet, whatever the object of our work, it's always the world of which we're thinking. It's all we have."

"And what will you play on your violin?"

"Something by Schubert, perhaps. An airy, melodious piece to relax me. Or some Bach. Something robust and supportive. I don't know, I haven't decided yet. I always decide at the last moment."

She asks him why, when he was so adventurous in his scientific breakthroughs, he has always remained attached to the music of former centuries. Why did he spurn Schoenberg, for example?

He doesn't know. That's how it is. He has never asked

himself that question. He has plenty of others to occupy him.

She thanks him again and expresses the hope that this won't be the last time they meet, but he says he isn't sure he'll be staying stay here much longer with Helen Dukas to look after him. He isn't even sure that his concepts, ideas, calculations and hypotheses still have a future. Knowledge is forever surpassing itself, he says again, or it would cease to be knowledge. The distant shapes are changeable, even when seen from on high.

"It's a mistake to claim, as people say I did, that what is incomprehensible is that the universe is comprehensible. One could even maintain the opposite, and it would be far closer to the truth: that what is comprehensible is that the universe is incomprehensible."

So it's ultimately incomprehensible. Nevertheless, he says, you have to go back to it again and again. It has to be tackled, this container of everything, regardless of the work involved, regardless of the repetition, effort and occasional tedium. Regardless of ordinary space, of time wasted and the oblivion forever lurking in the shadows.

"For all that," the girl says (consolingly?), "your theories have survived for a hundred years."

"What are a hundred years?"

She can't think of a reply, so she says: "Au revoir."

"Au revoir, young lady, and thank you for coming to see me."

He puts out his hand and she shakes it. Or rather, she tries to, but her hand passes straight through that of her host, who continues to smile at her. All she experiences is a faint, fleeting contact.

She takes a last look around. Albert Einstein gives her a friendly wave, then turns and goes over to his violin, which he picks up and briefly polishes on the sleeve of his sweater. He selects a piece of music and puts it on the stand.

He's behaving as if the girl isn't there anymore. He doesn't look at her. He's alone now. His expression is serene, almost grave.

All around him, after the universal chaos created by the hurricane sweeping through the doorway that opened onto the explosion, things have replaced themselves in a certain order. The papers and books are once more stacked on the desk and tables as if transported there by invisible hands. The pipes are back in their rack. The arrangement isn't exactly what it was, but the girl is inured to this imprecision.

She opens the door to the waiting room and slowly goes out, averting her gaze. Accompanied by the strains of the violin, she crosses the waiting room, which is now deserted. Where have all the people gone? Did they get tired of waiting? Will they return someday with their folders securely strapped?

She opens the front door, casts a last glance at the vacant, neatly arranged chairs. No sign of Helen, the woman who greeted her on arrival.

She goes out and shuts the door behind her. It really seems she can depart without difficulty, if not without regret.

With one hand on the banister, she descends the ill-lit

staircase. She can still hear the violin, but more and more faintly.

She makes her way along the hall and out into the street, which is almost deserted.

It's nightfall. Everything outside looks normal and peaceful. The first stars are already twinkling in the sky. A tram goes past, bell clanging. It bears the number 17 and is yellow and black in color, the girl notices. She stares after it for a moment, then looks at her watch. The second hand is moving again.

Everything is resuming its outward appearance.

Will there be anything on the tape recorder? We will never know.

The girl takes a few steps and looks up at a window on the first floor. We can no longer hear the sound of Einstein's violin. A light goes on, illuminating the window. Is that the room we were in? No. The study didn't have a window, the girl remembers, nor did the waiting room.

More lights come on at the front of the building—unsurprisingly, since the day is ending. Television sets go on too.

The girl strolls off down the street.

Turning for another look at the illuminated window, she fancies that up there on the first floor, in the big room with five doors, Einstein must have laid aside his violin. She pictures him pacing up and down, twisting a strand of white hair around his finger, going over to the blackboard, picking up a stick of chalk and swiftly jotting down some symbols. In a word, working.

- ACKNOWLEDGMENTS -

MY THANKS TO all the physicists and astrophysicists who accompanied me into the invisible: to Jean Audouze, Thibault Damour and, above all, to Michel Cassé, who read the draft of this book with a friend's exacting eye.